EASY
ECOLOGY

FOR THE BUSY BEE

BOB WARDINSKI

Published by Orange Hat Publishing 2019

ISBN 0-0000000-0-0

www.orangehatpublishing.com

♫ *This is dedicated to the one I love…*♫
To my wife Jeanette, for all her support, guidance, suggested ideas and
concepts that helped make this book possible.
THANK YOU!

TABLE OF CONTENTS

INTRO

Why this type of Ecology
book and its design

(Can you find the "Martian?")

I hope this book is totally unlike a text book and I hope the reader will find it more like a learning conversation. The book's layout is logically ordered, starting with basic science terms and concepts, leading into an integrated process of looking at our environment. It is design to hold the reader's interest while learning ecological concepts as it pertains to our home, planet Earth.

My goal is to keep the material as simple and direct as possible; interesting as well as practical for the average learner. Learning about our environment catches most people's interest at an early age as kids, only to have these interests fall back on the list of more important issues of their lives as they move into adolescence.

However, for many people, the interest in our ecology and nature reawakens as they learn more about life and become adults. Busy adults. People who have jobs, kids, and are taking college classes to advance their careers, never having the time to service their personal curiosities or intellectual needs about the workings of our natural world. Consequently, I noticed people who have this casual, or even deep interest in the Earth's ecology would have a hard time finding a resource that would meet their curiosity level, as well as their lifestyle. In addition, this type of book would need an added touch; Personality. I've always felt that personality, or the writer's voice, was the added touch that holds the attention of the reader, especially on topics like this one. Consequently, I hope that one will hear my voice in the pages that follow. However, the reader will need to appreciate my humor, or lack thereof, as well as my quotes from cartoons, music, and movies. Why? Because that's my personality. I want this book to be for all ages and learning levels. I especially wrote it in this fashion to attract an array of individuals who have that curiosity about the workings of our environment and the ways we've started to work with nature instead of against it. I frequently try to remind people that learning is a life-long process. It really begins at birth and continues to our deaths. Yes, everyone dies and pays taxes!

Now, I'm retired. Looking back, I still have this passion to pass on information. Plus, in my heart, I still love to learn. So, here I was retired, with all my "cool" overheads of notes and charts that I felt were too good to go right into the landfill, and I realized I still had a story to tell. Oh, for those of who may have forgotten, or perhaps never heard of an overhead projector it's a small unit that sat on a cart, or table top, at the front of classroom to display information being presented. The actual "overheads" were thin 8.5 x 11inch clear plastic sheets that you could write on. All types of teachers used them to write out fancy notes, display charts and graphs or cartoons like some of my own, including the Bizarro ones I chose for the whole class to see. That is why you will see them in the book. This is the personality and enthusiasm I hope my readers feel in these pages.

As I stated before, to make this book unique, yet informative, I wanted my notes to tell the story of how our ecology works. When possible, and in close proximity to the formal text of this book, one will notice the fantastic and most interesting notes I used in my teaching style. However, in the transferring of my notes to these pages it seemed to have made them a little "scruffy" looking, but in the end, I saw it as a positive look, making the pages look more personal. In the formal text I will go over the notes to which I'll refer and provide further examples and thoughts on the materials being covered. In addition, I numbered my notes. For example, one will see the "Notes and Number," (as well as the Page Number) for the symbol in my discussion that coincides with the information being covered for the reader to refer to as they move through the discussion. Also, as I move on to other sections I can reference the reader back to previous notes and their numbers, for further clarification or examples of topics already covered. I hope everyone enjoys my discussions and information provided, perhaps even to the point of going out and researching some of the topics covered.

Ok, it's time to get started! I hope I've gained the attention of anyone who has even the smallest curiosity about how our ecology works. Those of you who want to know what global warming is all about, why landfills are wrong, why pollution and fossil fuels are integrated roles in our life, as well as the consequences thereof. I want every newfound learner to read this book, to learn the most basic laws of nature, energy, and mass, and how they impact the characteristics of life on our planet. I want everyone to know what plants have to do with it, what makes a muskrat guard its musk, or, as Tina Turner sang, ♫ *What's Love Got To Do With It* ♫ (1984).

Let's find out!

SECTION ONE

The Beginning, a Review of Ecological Studies

(Can you find the "Eye Ball?")

Many people said that as a teacher I was thorough. I provided a logical beginning and worked my way through each unit with a logical conclusion. In the past I would tell my students I would never leave parts out of a story for them to piece meal together later with their imaginations. And, I continue with that thought to this day. This was important to me, because it happened so often with my education. I would use the example of a movie trilogy. I would never start a trilogy at part two, never assume they saw part one, and then never show part three.

Consequently, I never take anything for granted so I'll begin this section by defining the term science - the study of a subject. (Notes 1, page 14) Notice the number 1 in the upper left-hand corner of my notes on page 14.

Generally, most people would think of defining science as a term used only as it relates to biology or chemistry concepts. This is not the case. The truth of the matter is that one can make a science out of anything. A good example would be myself. As a kid I made a science out of Tiddly Winks. Yes, I would study the distances and how much pressure to place on my chips so they would flip and go right into the cup. Then I would get "the look" from my students, perhaps the same one my reader is having right now. I call it the minion look: "Whaaaaat?"

1

WHAT IS

SCIENCE ?

↳ THE STUDY OF A SUBJECT

EX.) – POLITICAL SCIENCE
– POLICE   SCIENCE
– EARTH   SCIENCE

– ENVIRONMENTAL SCIENCE

↳ THE STUDY OF THE IMPACT OF HUMANS ON THE ENVIRONMENT

"What are Tiddly Winks?" I would try to explain the game, but in the end it's best if you just go online and look it up. Again, whether it be cards, shooting pool, or roasting corn, one can make a science out of anything. In either case, as my notes show, there are many sciences that have developed over time. For example, when scholars began to take note of and document the impacts that human populations were having on the environment, they started to study these problems. We now know this as environmental science. (Notes 1, page 14)

There is another concept I would stress to my reader right from the beginning. It's what I call the "defining words as the prefix-suffix" game. (Notes 2, page 16) I would call it a game because that's what it is, a game, and who doesn't like a game as a way of learning. It's a process of combining prefixes and suffixes to create new words. I found out early that vocabulary is a big part of learning any science. It only stands to reason, "if one were to be going to Spain, wouldn't it be nice to speak Spanish?" It really helps if one knows the language, or vocabulary, and it would be no different when studying the foreign lands of biology, ecology, or any other science. One needs to know the vocabulary.

2

COOL

WHAT ABOUT ME?

WHAT IS ↳ BIOLOGY ?

LIVING   STUDY OF...

ECOLOGY → SEEKS TO UNDERSTAND THE RELATIONSHIPS AND INTERACTIONS BETWEEN ORGANISMS AND THEIR ENVIRONMENT

GREEK FOR "HOME"   STUDY OF

* BRANCH OF "BIOLOGY"
→ OVERLAPS W/ GENETICS

BIOCHEMISTRY
* EVOLUTION
PHYSIOLOGY

As I show in my notes, I would remind you to always break down the word. For example, ecology can be broken up into "eco" (Greek for home) and "ology" (the study of . . .). (Notes 2, page16) Consequently, I would suggest all my readers go online and print a list of the most widely used prefixes and suffixes and their meanings. Not only are these lists readily available but are a great resource for learning the vocabulary of any subject (Note 2, page16). Also, in my notes, I added a list of symbols and/or abbreviations the reader of this book would need to become familiar with to study and understand the concepts presented in ecology (Notes 3, page 18). Some are official symbols used in the biological, chemical, and other related ecological sciences. Others are of my own doing. My philosophy has always been, "don't fight it, enjoy it." So, I made up symbols for critters and other related items to keep it fun as well as interesting. Check out my Notes 3, page18. I found that for many people, it was hard to visualize some critters that are important to our ecology. For example, bacteria are microscopic organisms that play a big role in our environment, but are of difficult to visualize, so I created a simple Packman-looking symbol to represent them. It's easy for people to remember, it's fun, and it gets the job done. The prefixes and suffixes of words I'll provide in the notes and I'll try to define the most necessary as they relate to ecology.

SYMBOLS TO REFER TO

LOOK

= Look at my notes

E = Energy

e^- = Energy flow

☀ = Sun or Solar Energy

↑ = Given off (ex. Gas, Heat)

= Fungi

= Bacteria's

C – C = Carbon Bonds

= Arrow into living things

= Arrow out of living things

= Human Species

Plants or Producers =

↑ = Increase or Gas given Off

↓ = Decrease or Precipitate

Land = _____ OR

C = Carbon

N = Nitrogen

Fe = Iron

O = Oxygen

CO_2 = Carbon Dioxide

P = Phosphorus

PO_4 = Phosphate

= Atmosphere

= Water

= H2O or Water cycle

SPACESHIP EARTH

OUR HOME (CAPSULE) FLOATING IN SPACE

CLOSED SYSTEM

NOTHING IN ↔ NOTHING OUT

" PALE BLUE DOT..."

CARL SAGON
↳ ASTRONOMER
- PHYSICIST
- BIOLOGIST
- AUTHOR

As one would find out we will be studying our home, the Earth. The one and only place we can live. As the late and famous Carl Sagan once said, the earth is like a space ship, a capsule floating in space. The third rock from the sun. All the history we know happened right here on the Pale Blue Dot. (Sagan,1994) A closed system; everything we need and have is right here – nothing in, nothing out (Notes 4, page 19). Put simply, we would be studying the relationships between the living (biotic) and non-living (abiotic) portions of our biosphere that had, over immense amounts of time, developed into many unique ecosystems. (Notes 5, page 21). Through evolutionary events, as organisms became reproductively isolated, species-specific populations emerged (Notes 6, page 22). Through time, these communities developed relationships that mostly helped each species survive and progress. In some cases, the relationships proved to be detrimental and some species perhaps didn't fare as well and were lost. Ultimately, the communities with favorable biotic and abiotic relationships gave rise to what we now call the biosphere. The living round. The three main components of our thin layer of life are the land, atmosphere, and water, all of which work closely together. (Notes 7, page 24)

ECOSYSTEM — AN INTERACTING SYSTEM
OF GROUPS OF ORGANISMS
LIVING TOGETHER W/ THEIR
NON-LIVING OR PHYSICAL
ENVIRONMENT

OR

LIVING THINGS ⟶ INTERACTING ⟶ NON-LIVING
WITH   THINGS

BIOTIC
LIVING PARTS
OF THE ENVIR.

EX.) ANIMALS
PLANTS
MONERA (BACTERIAS)
FUNGI
PROTIST

ABIOTIC
W/O LIFE
NON-LIVING
PART OF THE ENVIR.

EX.) H_2O
LIGHT
TEMP.
INORGANIC
SUBSTANCES
(PERIODIC
TABLE)

6 THE FLOW OF LIFE IN OUR HOME

ORGANISM

— AN INDIVIDUAL LIVING THING

SPECIES — GROUP OF ORGANISMS THAT ARE REPRODUCTIVELY ISOLATED

* CAN ONLY HAVE OFFSPRING W/ EACH OTHER

POPULATIONS

— ARE ALL THE MEMBERS OF THE SAME SPECIES LIVING IN THE SAME PLACE AND TIME

COMMUNITY

— A GROUP OF VARIOUS SPECIES THAT LIVE IN THE SAME PLACE AND INTERACT WITH EACH OTHER

If one looks at the diagram showing the impacts of climate, soils, plants, plus other life forms, including humans, one would have to agree they're all connected (Notes 8 page 25). The soils wouldn't have formed without the climate's weathering process (which I'll discuss later). The plants wouldn't have been able to grow and expand without the soils to provide the nutrients. And, of course, other critters couldn't exist without the nutrients from plants. And, I didn't even include the flow of gases, liquid water, and solar energy that are required by all life forms. (Notes 8, page 25)

Whether we are studying a lake, prairie, farm field, or ocean, the end results are the same. All ecosystems share certain structural and functional characteristics that make them uniquely connected. (Notes 9, page 27) What I mean by ecosystem structure is really how it is designed; how different portions of our environment interact the way they do. For example, and for simplicity's sake, let's say oak trees produce acorns only for squirrels and that hawks feed only on squirrels. Why does the oak tree produce so many acorns? Well, in quantity, the tree increases its chances of passing on its traits to potential offspring by way of its seeds (the acorn). It's not the traits that are of interest to the squirrel but rather the treats that are in the acorns. The structure is the tree that picks up the nutrients from the surrounding soil and atmosphere, and with the help of the sun, packs them full of goodies. That's why the squirrel goes after and eats the acorns. For the very same reason the hawk comes down and eats the squirrel.

7

ECOSYSTEM

— ORGANISMS INTERACTING

WITH THE LIVING AND NONLIVING
    (BIOTIC)   (ABIOTIC)

PORTIONS OF THE ENIVRON.

O_2

CO_2

CO_2

BIOSPHERE

LIVING

— THE THIN LAYER OF EARTH
WHERE LIFE EXISTS

<u>3 PARTS</u>

1.) LAND

2.) ATMOSPHERE

3.) WATER

OUR HOME

✳ ALL 3
AREAS WHERE LIFE IS
FOUND

8

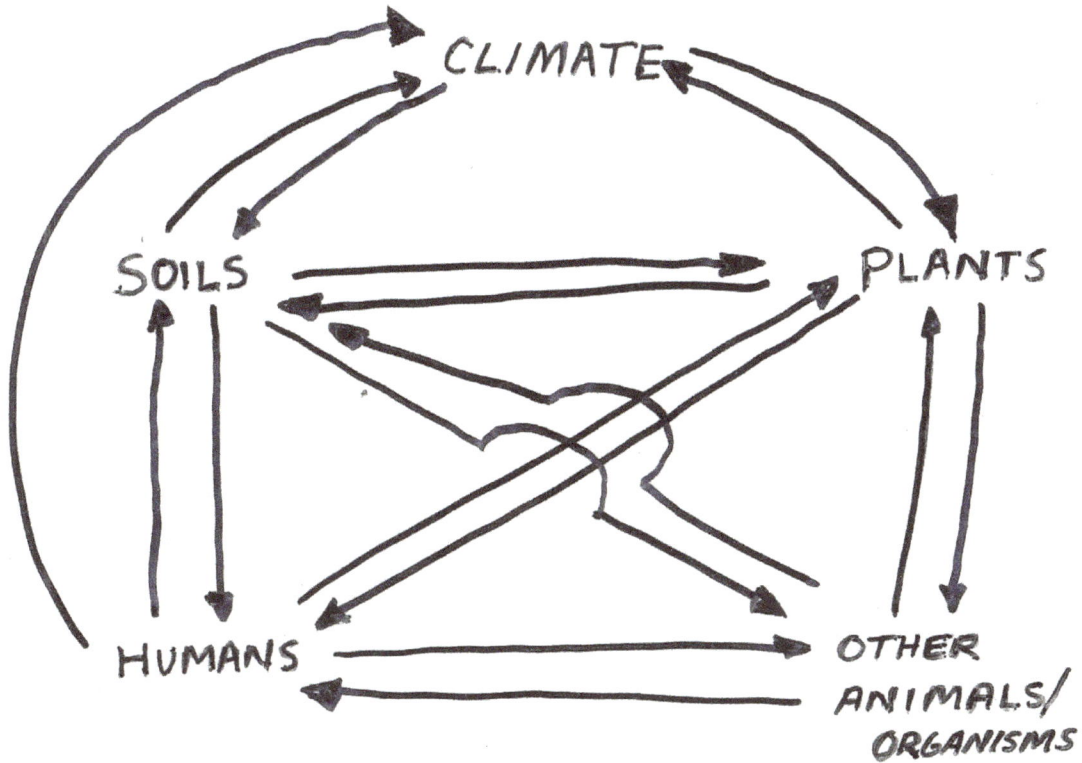

MOST IMPORTANT

AN ECOSYSTEM'S VARIOUS COMPONENTS ARE
" HIGHLY INTERRELATED

BIOTIC OR ABIOTIC → EACH INFLUENCES
THE OTHER

* ALTERING ONE COMPONENT
WILL ULTIMATELY AFFECT
ALL THE OTHERS

" ALL THINGS ARE CONNECTED... "

The tree benefitted by having its genetic traits dispersed by way of the squirrel who tends to eat some and bury some for later. The squirrel benefits from the nutrients and energy provided by the acorn which was ultimately packed in the seed to grow into a tree. Plus, the hawk now benefits by way of eating the squirrel to gain its nutrients and energy; all because of one oak tree. This is a little snap shot of the many interactions that occur in our biosphere known as "ecosystem structure." (Notes 9, page 27)

Ultimately, if the oak tree wasn't there, the squirrel couldn't eat the acorns and the hawk couldn't feed on the squirrel. The result is both would die because they are dependent on the oak tree. This, of course, is a simplified example of an ecosystem structure, but the net result is still the same. Even with more complex interacting systems, it allows life to continue. Looking at the human side of ecosystem requirements, it's interesting to note if one goes back and looks at my Impact flow chart (Notes 8, page 25), if one were to remove the portion of the "human impact" the system would still function. However, remove any of the others and humans might as well get out their ten-pound bag of sugar and sit on it.

9 | ECOSYSTEM : STRUCTURE/FUNCTION

STRUCTURE   VS.   FUNCTION

- CONSTRUCTION OR DESIGN

- WHAT IT DOES OR WILL DO
- USES

* CAN TELL US ABOUT HOW IT FUNCTIONS

* CAN TELL US ABOUT ITS STRUCTURE

STRUCTURE OF AN ECOSYSTEM

- IS THE RELATIONSHIPS BETWEEN ORGANISMS AND THEIR NONLIVING ENVIRON.

FUNCTION OF AN ECOSYSTEM

- IS THE ABILITY FOR LIFE TO EXIST

Now you're giving me that Minion look. "Whaaaat?!" My reply--Why? Because they can kiss their sweet butts "good-bye."

For most organisms, the sole goal in life is to eat and reproduce (except for humans). With all the interactions that have evolved, it should become obvious the main function of any ecosystem is to allow life to exist. (Notes 9, page 27) With that in mind, there are three basic requirements for all ecosystems in our biosphere. Whether it be on land, in the water, or in our atmosphere, all require (1) the cycling of nutrients, (2) the flow of energy, and (3) structure, also known as predator-prey relationships. (Notes 10, page 29) The ecological structure, mentioned earlier, with all of its interactions, supplies the opportunity to meet the needs of the first two requirements. In this instance, just think of oneself. Why does one eat? Ok, one gets hungry. However, the reason for a growling stomach is the body is calling for nutrients and energy. Simultaneously, the body is extracting energy from the food either to be used immediately or stored for later use. Also, think about when you are eating. You are engaging in a predator-prey relationship of sorts - ecological structure. The key importance here is, and will always be, as we move forward, the thoughts of Chief Seattle 1854 (Notes 11, page 30). "All things are connected. . .like the blood that unites one's family. . .We did not weave the web of life, we are merely strands in it. . .what we do to the web, we do to ourselves. . ." (Seattle,1991)

10 ALL ECOSYSTEMS SHARE CERTAIN COMMON STRUCTURAL / FUNCTIONAL CHARACTERISTICS

3 BASIC REQUIREMENTS OF ALL ECOSYSTEMS

1.) ALL REQUIRE THE CYCLING OF NUTRIENTS

2.) ALL REQUIRE ENERGY (E)

3.) ALL REQUIRE "STRUCTURE"
ALSO KNOWN AS
"PREDATOR PREY RELATIONSHIPS"

11 THIS IS A SHORT EXCERPT/PARAPHRASED
FROM THE QUOTES OF CHIEF SEATTLE
(1854)

THIS WE KNOW:...THE EARTH DOES NOT
BELONG TO US; WE BELONG TO THE EARTH

THIS WE KNOW: ALL THINGS ARE CONNECTED
LIKE THE BLOOD WHICH UNITES ONES
FAMILY, ALL THINGS ARE CONNECTED.

WHATEVER BEFALLS THE EARTH BEFALLS
THE SONS AND DAUGHTERS OF THE EARTH.

WE DID NOT WEAVE THE WEB OF LIFE,
WE ARE MERELY STRANDS IN IT.
WHATEVER WE DO TO THE WEB
WE DO TO OURSELVES...

This is a very important quote. It should act as a reminder of all the impacts we have had and continue to have on earth. Altering one component will ultimately affect the others. With that in mind, as I move into the next three sections of nutrient cycling, energy flow, and predator-prey relationships, I'll finish each with some examples of the human impacts and/or practical applications of why they are important to know.

Again, everyone needs to keep in mind how closely we are attached, directly or indirectly, to all living things in our biosphere. The way we live, the decisions we make, and the impacts those decisions have will play a big role on the quality of life for future generations. Remember, as Chief Seattle said "…We are all connected. . ." (1854)

At this time, it's only fitting that I should have you pull out your copy of, ♫ *Circle of Life* ♫ by Elton John (1994) and play it. You do have one? If not GET ONE! It is a beautiful song.

SECTION TWO

Nutrient Cycles
Groups of Players,
and Terms

(Can you find the "Slice of pie")

Nutrient cycling, as I mentioned before, is the first of three requirements for all ecosystems. Even though I will cover each of these requirements individually, we must remember they are all working together simultaneously for life to exist. Also, there are groups of organisms, or types of players, that are fundamental to all three of the ecosystem requirements. (Notes 12, page 36) It's important to know these groups of individuals by name and what role they play in the biosphere. These are the most basic groups of organisms that provide the connections for all living things. I like to think of these individuals as groups of helpers that set the kitchen table for each other to feast upon.

The first group of important players is the "Producers." (Notes 13, page 37) These are the organisms that begin the cycling of nutrients through the ecosystem. When one sees my flower symbol (Notes 3, page 18) one needs to think of all the types of organisms that carry on photosynthesis. One needs to think of the varieties of trees, grasses, algae, and microscopic planktons that have this unique ability to use the sun's energy, photo, or light, to make or synthesize carbon compounds into living tissues.

SECTION TWO: NUTRIENT CYCLES

YUM

REMEMBER: THE 3 BASIC REQUIREMENTS
OF ALL ECOSYSTEMS

1.) NUTRIENT CYCLING

2.) ENERGY FLOW (E)

3.) STRUCTURE — INTERACTIONS
OR PREDATOR PREY
RELATIONS

GROUPS OF PLAYERS (OR ORGANISMS)
THAT ARE FOUND IN ALL 3 ECOSYSTEM
REQUIRERMENTS

HELLO

Hi

Hi

Hi

Hi

PRODUCERS

- ORGANISMS THAT PRODUCE THEIR OWN FOOD

✻ MAINLY PLANTS

→ USE "CHLOROPHYLL" TO

CARRY ON PHOTO SYNTHESIS
LIGHT   TO MAKE.

$$CO_2 + H_2O \longrightarrow ORGANIC\ COMPOUNDS$$
SUN "E"   EX.) TREES
                                APPLES

AUTOTROPHS
SELF   FEEDING

IN OTHER WORDS THESE GUYS UTILIZE THE SUNS "E" TO GROW AND REPRODUCE

NOTE ✻ THERE ARE OTHERS
        EX. PROTIST

 — HI

Additionally, by making these tissues made of chemical bonds of carbon, the producers store the energy for all other living things. This is a process I'll explain in more detail in the section on energy. Producers are also known as "Autotrophs" because they can harness the sun's energy to grow and reproduce, regardless if they're a plant or protist, etc. In other words, they can feed themselves, (i.e., self-feeders).

Next in the series of important organisms, are the "Consumers." (Notes 14, page 39) These guys cannot make their own food and are dependent on "Consuming" (or eating) plants and/or other animals to meet their nutrient and energy needs. For these reasons, consumers are also known as "Heterotrophs." This is a wide variety of organisms that feed on (or consume), directly or indirectly, the tissues of producers. For example, "Herbivores" feed on microscopic phytoplankton and other plant vegetation. Examples of herbivores include deer, cows, and squirrels, just to name a few. Then, there are the higher-level consumers that like to dine on plant vegetation, for instance, tree seeds, berries, and leaves but enjoy a meat-based meal when available. We call these consumers "Omnivores." There are many examples of these, such as rodents, bears, skunks, and, oh yes, "Humans." That would be if one is not a vegetarian! Next are the "Carnivores." For the most part, these are critters that are dependent on eating meat to meet their nutrient and energy needs, such as tigers, eagles, and muskies. (Notes 15, page 40)

14   <u>CONSUMERS</u>

- ORGANISMS THAT CANNOT
  PRODUCE THEIR OWN FOOD
  BUT OBTAIN "E" FROM THE
  FOOD THEY <u>EAT</u>
           CONSUME

- FEED DIRECTLY OR INDIRECTLY ON
  PLANTS

→ <u>HETEROTROPHS</u>
   DIFFERENT   FEEDING

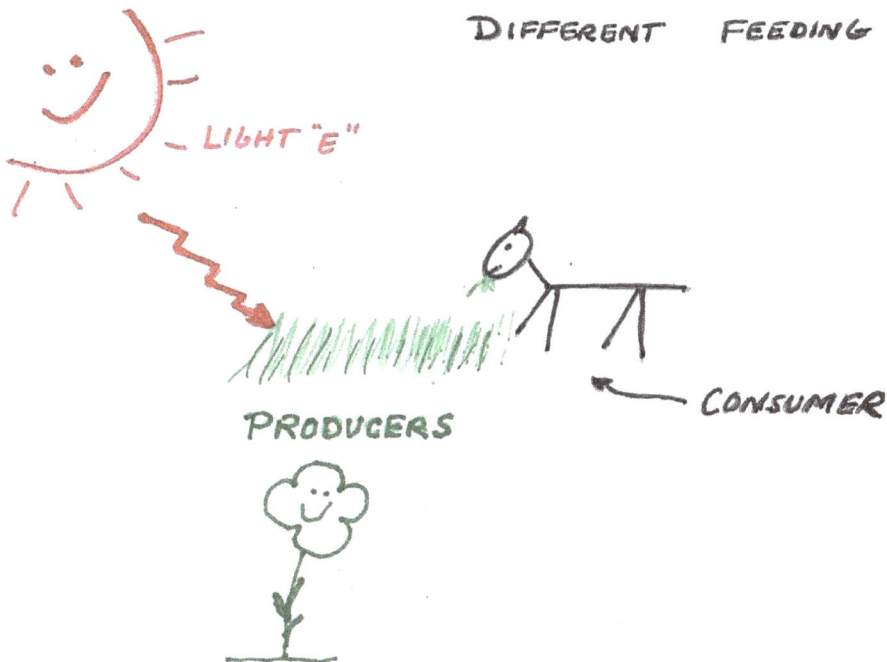

LIGHT "E"

PRODUCERS

CONSUMER

TYPES OF CONSUMERS

HERBIVORES
→ EATS ONLY PRODUCERS
OR "PLANT EATERS"

OMNIVORE
→ EATS BOTH PLANTS AND
ANIMALS

CARNIVORE
→ EATS ONLY OTHER CONSUMERS
"MEAT EATER"

DECOMPOSERS
→ WHAT'S ON THE PLATE?

BREAKS DOWN DEAD ORGANISMS
CAUSING THEM TO ROT

EX.) BACTERIA & FUNGI ← MAINLY *THESE TWO*

SOME
ANIMALS — EARTH WORMS
PROTIST

To demonstrate how ecosystems set up their structure, or interconnections, would be to look at a simple scenario of feeding levels. The grass grows (producers) and the deer comes along and eats the grass (herbivores). In turn, the human goes hunting, kills the deer to eat alongside a nice craeser salad (omnivores). The human then decides to go swimming in the ocean and is eaten by a shark (carnivores). YIKES! Oh, yes, I forgot to mention that humans are not on the top of the food chain, a topic I'll discuss in more detail later.

Finally, there is the greatest of all consumers known as "Decomposers." Without these guys nutrients from the wastes of other lifeforms, as well as their tissues, would never return to the biosphere for reuse. One can only imagine what our forests would be like if trees, with their branches and leaves, didn't breakdown and return their potential nutrients back into the soil. First, we would have dead trees and their tissues piling up over time. Secondly, the trees eventually wouldn't be able to even grow because the soils in which they grow would be depleted of most of the necessary nutrients that had been taken up and stored in the tissues of the dead organisms. (Notes 15, page 40)

#1 NUTRIENT CYCLES

NUTRIENT

- AN ESSENTIAL ELEMENT OR CHEMICAL THAT CAN BE ABSORBED AND USED BY AN ORGANISM

CYCLE

- REUSED OVER AND OVER

INORGANIC NUTRIENTS

- COMES FROM ELEMENTS OTHER THAN CARBON

  Ex.) - POTASSIUM (K)   NITROGEN (N)

  - PHOSPHORUS (P)
  - ZINC (Zn)
  - H_2O , O_2

ORGANIC NUTRIENTS

- COMES FROM CARBON COMPOUNDS

  Ex.) ORGANISMS $\longrightarrow$ LIVING THINGS

  ORGANIC

2 TYPES OF NUTRIENT NEEDS

17

MICRONUTRIENTS

SMALL   — NEEDED IN SMALL AMOUNTS

           EX.) PRODUCERS USE SMALL AMOUNTS OF IRON, MAGANESE, ETC.

MACRONUTRIENTS

LARGE   — THOSE NEEDED IN LARGE AMOUNTS

           EX.) PRODUCERS USE LARGE AMOUNTS OF NITROGEN, PHOSPHORUS, ETC.

Nutrient cycling, as I mentioned earlier, is a very important system of moving inorganic and organic nutrients through the biosphere. (Notes 16, page 42) Nutrients are essential elements and/or compounds that are required for life to exist. To set the record straight, when one reads the term "inorganic nutrients," it really means it comes from elements, or compounds, of non-living sources. For instance, rocks and soil particles release Calcium, Magnesium, Potassium, and Zinc, all of which originate from non-living, noncarbon-based tissues. Carbon is a unique element in our biosphere and will be given its much-deserved attention when I discuss the Carbon cycle. But, yes, there are inorganic forms of carbon. I would always ask how many of my students love diamonds. I would tell them diamonds are pure carbon in its crystallin form. Graphite is another inorganic form of carbon. In this form, the carbon structures are very small particles that take on a more roundish shape. This allows the particles of inorganic carbon to roll upon each other and makes for good writing utensils like pencils. Or, graphite, with its rolling structures, can act as a very good dry lubricant. It is often used on padlocks to prevent them from seizing up over time.

Anything that is alive, or is dead and rotting, is made up of organic-based carbon. Organic comes from the word organism, or living things. Microscopic critters, skin, hair, fur, lions, tigers, and bears – "OH MY," are all good examples of organic/carbon-based tissues of living things. (Notes 16, page 42) All organisms, from producers to consumers, require organic and inorganic forms of nutrients; the only differences are based on their individual types and quantities they need. For example, plants require some nutrients in small amounts, like iron, while those same plants need large amounts of Nitrogen to survive. The nutrients needed in small amounts are called "Micro-nutrients," while the nitrogen needed in large amounts is known as a "Macro-nutrient." The micro- and macronutrient needs of individuals varies as much as the wide range of organisms that live on Earth. (Notes 17, page 43)

18

THE 4 MAIN NUTRIENT CYCLES WE WILL BE LOOKING AT

A.) MINERAL CYCLE

B.) CARBON CYCLE

C.) NITROGEN CYCLE

D.) PHOSPHORUS CYCLE

NOTE* ALL ELEMENTS ON THE PERIODIC TABLE ARE IMPORTANT TO LIFE FORMS

* OUR FOCUS WILL BE ON THESE 4 NUTRIENT CYCLES

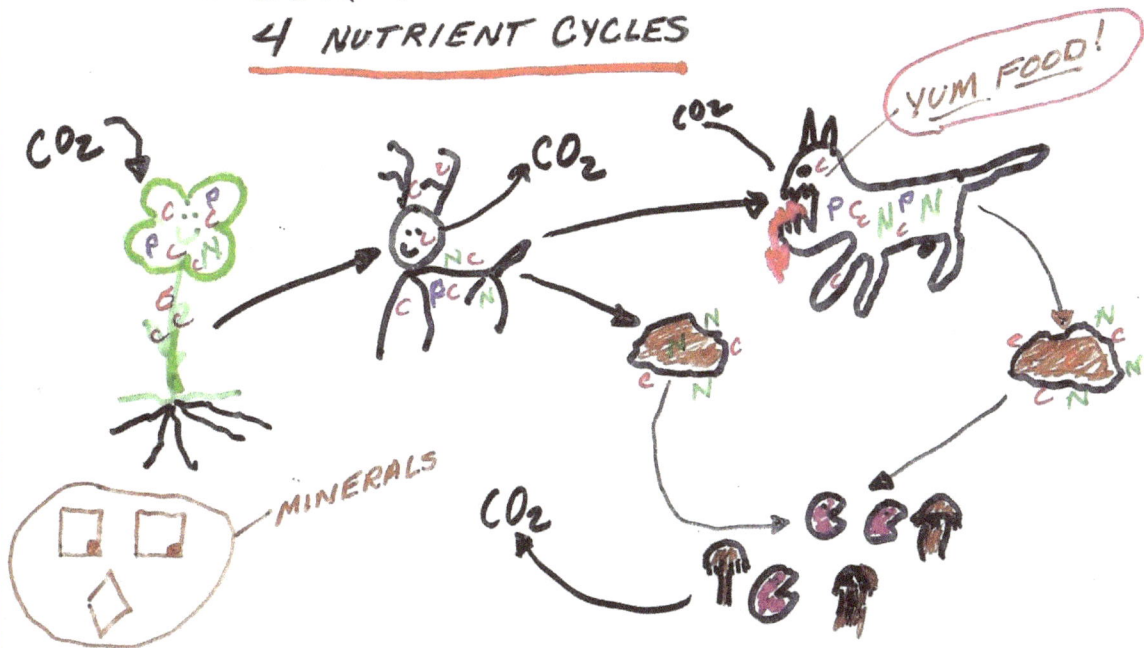

The four main nutrient cycles I'll cover in this section will be:

(A) The Mineral Cycle

(B) The Carbon Cycle

(C) The Nitrogen Cycle

(D) The Phosphorus Cycle.

(Notes 18, page 45)

Remember, most elements on the periodic table are important to some degree as their needs vary. Nonetheless, the ecological and environmental sciences have designated these four nutrient cycles to be the most important in our biosphere because of their great necessity for all living things. Plus, these four nutrient cycles and the organisms involved play a very important part in the impact humans are having on our planet. (Notes 18, page 45) Remember Chief Seattle's words "...the Earth does not belong to us; we belong to the Earth." At this point, I would usually tell my students it's time for an ecological song, so now I would ask my reader to look up a copy of, and listen to the song, ♫ *Mercy, Mercy Me, the Ecology* ♫ by Marvin Gay (1971). It's a hip song!

THE MINERAL CYCLE

A. MINERAL CYCLE

MINERALS = ELEMENTS →   } INORGANIC
          EX.) Ca, P, K,   (FROM NON-LIVING)

⇓

ORIGINATE FROM

ROCKS → RELEASE OF MINERALS
THROUGH A PROCESS KNOWN
AS "WEATHERING"

   — FREEZING / THAWING

   — HEATING / COOLING

  * EROSION FROM WIND AND
    WATER

  * VERY SLOW PROCESS

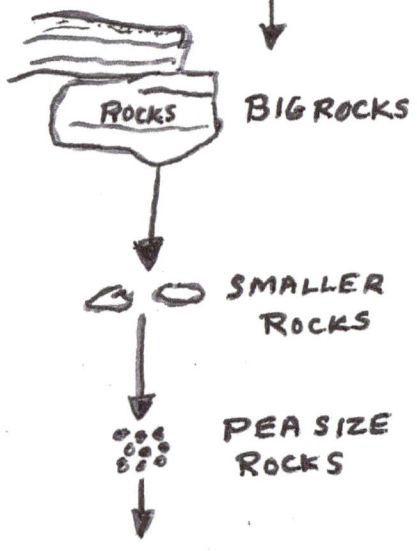

BIG ROCKS

SMALLER ROCKS

PEA SIZE ROCKS

SAND
(GRAINS)
— COURSE

SILT
SMALLER GRAINS
— TACKY

CLAY
SLIPPERY

} 3 MAIN COMPONENTS/ PARTICLES THAT MAKE UP SOILS

The mineral cycle begins with the releasing of elements and/or their compounds by way of a process known as "Weathering." This is a very slow process of breaking down rocks into fine particles that release their nutrient contents for producers and other critters to absorb. To understand how the weathering process works, one needs only to think of the great glaciers that covered large portions of the Earth. The glacial advances and retreats broke apart rocks, gouging and carving out many of the great geological formations we enjoy today. These gigantic sheets of ice, with their glacial mass rising (sometimes two miles high in thickness!) made many advances and retreats as they froze and thawed. This back and forth movement of the ice sheets pulverized and ground the Earth's rocks into ever smaller particles. Ultimately, these particles were released into the vast flows of melting water, there they would be tumbled and ground into even finer grains of sand, silt, and clay (Notes 19, page 48).

This process created the fertile soils we have in much of the United States and around the globe. Once these particles settled out and released their contents of elements and compounds, they also allowed plant-based autotrophs to move farther onto the land. As plants advanced onto the land, they grew, died, and decayed. This would begin the process of adding organic nutrients to the soil base.

PROGRESSION OF SOIL FORMATION

ROCKS
WEATHER

} MINERAL CYCLE
RELEASE THEIR
MINERALS BY WAY OF
SAND, SILT, AND CLAY

ALLOWS
PLANTS TO
GROW AND
DIE

PLANT GROWTH AND DECAY
ADDS ORGANICS TO
THE MINERALS

ALLOWS DECOMPOSERS
TO MOVE IN

      EX.) BACTERIA, FUNGI → BOTH ADD
ORGANICS TO THE
SOIL

HIGHER ORGANISMS
MOVE IN TO WORK./EAT/FEED
ON THE MINERALS AND
ORGANICS

      EX.) - MILLIPEDES

        - EARTH WORMS

The decaying process was the result of decomposers moving in to feast on the remains of plant tissues. The arrival of decomposers chowing down on rotting tissues, not only allowed these organisms to eat and reproduce, but also resulted in the releasing of inorganic and organic based elements and compounds back into the soil to aid in future development of other lifeforms. For example, the nutrients added during the decaying process furthered the growth of producers, which attracted and welcomed higher level, more developed, plants and animals. These new arrivals would begin to thrive by feeding, not just on plant tissues, but on the remains of micro- and macroscopic individuals of fungi, protist, and insects, consequently adding even more diversity to the surrounding ecosystems. (Notes 20, page 50)

The net result of the weathering process attracted and collectively added organic materials that, through time, would build and add further qualities to the soil. For instance, the buildup of humous (organic materials) was an added benefit by holding more moisture and keeping the soils loose for organisms to expand out and penetrate deep into the land (Notes 21, page 52).

21

BUILDS A RICH

<u>SOIL</u> — W/ <u>HUMUS</u> → ORGANIC
MATERIAL

— W/ MINERALS → ELEMENTS
FROM ROCKS
EX.) K, P, O, Si

THE MINERAL CYCLE

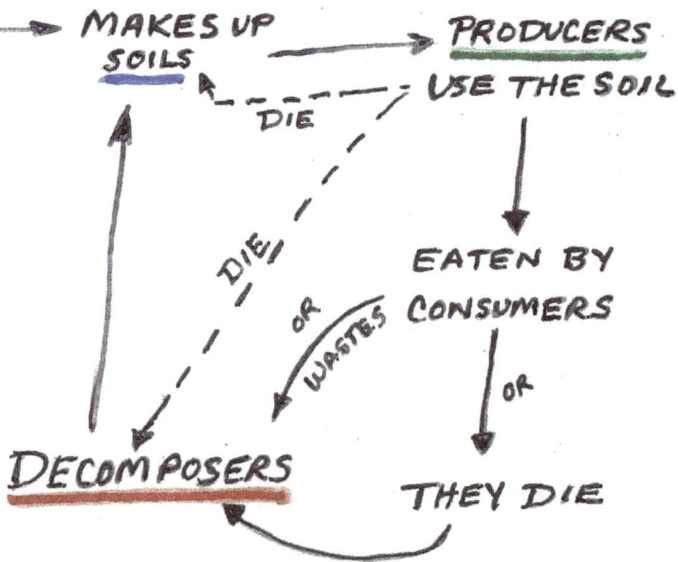

<u>ROCKS</u>

RELEASE ——→ MAKES UP ⤏ PRODUCERS
MINERALS   SOILS   USE THE SOIL
EX. Ca   ⟵ - - - DIE
    Mg
    S   DIE   EATEN BY
    Fe   OR   CONSUMERS
                         WASTES   OR

DECOMPOSERS   THEY DIE

MINERAL CYCLING CAN BE

FAST ⟶ W/IN HOURS

      EX.) THE USE OF O_2 AND CO_2 IN RESPIRATION

SLOW ⟶ LONG-TERM

      EX.) TIED UP IN ORGANISMS BURIED DEEP IN THE GROUND

          OIL OR COAL OR ROCKS

✳ **WITH ALL NUTRIENT CYCLES**

⟶ LAW OF CONSERVATION OF MASS

— NO CHANGE IN THE TOTAL MASS OF A SUBSTANCE INVOLVED IN A CHEMICAL CHANGE

WHAT IS USED IS ALWAYS ⟶ THERE UNTIL IT'S RELEASED

(START WITH) = (FINISH WITH)

EX.)

FLOWER PICKS UP 10g OF CARBON DURING ITS LIFE ⟶ IT WILL RELEASE 10g OF CARBON BACK TO THE ECOSYSTEM

☺ REMEMBER "A CLOSED SYSTEM"

A prime example of these lands is found right here in the midwestern portion of the United States, also known as the bread basket of the world because of its ability to produce large amounts of grains from its fertile soils. (Faith, n.d.) The mineral cycle is briefly laid out in my Notes 21 on page 52. When looking at this diagram, it's important to remember that the producers and consumers are assumed to range from microscopic to full-fledged animals. For example, producers can range from some of the smallest organisms, such as plankton, to full-fledged redwood trees that are among some of the largest and oldest organisms on our planet. Needless to say, all living things at some point eventually die and return their minerals back to the soil. The mineral cycle can be fast like the uptake of oxygen and the release of carbon dioxide from cellular respiration. Or, it can be slow; for instance, some minerals can be tied up in rocks or fossil fuels for immense amounts of time. Fossil fuels are dead plants and animals that were mostly trapped underground, decomposing under extreme temperatures and pressures for thousands of years. Trapped until we started to unearth them for their energy and by-products. This topic will be covered in more detail in later sections.

This leads me to one of the laws of nature that most people don't think about, especially when they see their garbage go into trucks and drive away. I would ask my students if they ever heard of the saying, "out of sight, out of mind," and most would raise their hands. I would tell them "that stuff is still with us." It might not be in their garbage cans or even in the neighborhood, but it's still there. It's called the "Law of Conservation of Mass" (Notes 22, page 53). It's an overlooked fact of nature and chemistry that states that there is no change in the mass of a substance involved in a chemical reaction. (Petrucci, 1989) In simpler terms, the definition explains that whatever one starts with, one has to finish with. "In has to equal out." Reactants have to equal products. Perhaps you remember, from your past physical science or chemistry classes, having to use coefficients to balance reactants with the products for some of those chemical reactions that had to be written out. That's because of the Law of Conservation of Mass. Being a law of chemistry makes it a law even on the ecological level. Whatever one starts with, one finishes with (Notes 22, page 53). For example, if a flower, during the process of photosynthesis picks up 10 grams of carbon during its life, at some point those 10 grams of carbon are going to be released back into the environment. Yes, even the elements that made up our forefathers, who were embalmed, placed in a sealed casket, and placed in the ground surrounded by a cement crypt, will eventually return to the earth. It's just a matter of time. Remember "…We are part of the Earth and it is part of us…" (Chief Seattle, 1854)

"The way I see it, we're all eventually dust in the wind. Oh, by the way that's a great song! Yes, it's really a song; ♫*Dust in the Wind*♫ by Kansas (1977). Try it – you'll like it!

The Mineral Cycle: Human Impacts and Recycling

Fast, slow, or anywhere in between, one only has to remember the natural processes of weathering, decomposition, and/or other consumers, not to mention possible cataclysmic events; that is, all routes that inevitably return all living things back to the earth for future critters to be reused. Remember, "in" equals "out." Whatever we start with, in the end, it's still with us.

The recycling of elements, compounds, and the tissues of dead organisms, is nature's way of balancing the "ins and outs" of nutrients that circulate through the biosphere. This is the natural process by which nature uses and recycles materials, a concept human's need to be more aware of with the production of products and their eventual disposal making them more, if not totally, recyclable. Consider just two items, glass and metals, that are thrown away every day. These are two items that can readily be recycled. If separated and melted down, both can be easily formed into new cans and bottles. For example, aluminum cans can be melted, reformed, and back on the shelves in less than six to eight weeks. (Why Aluminum Cans, n.d.) It really doesn't make sense to dig big holes in the Earth to extract raw ores in order to produce glass and metal products that are then used, discarded, and eventually thrown away into massive mounds called "landfills." It defies logic.

Even with recycling programs now in place, or placing a monitory value on these items, there are still many potential resources that are being wasted. Think about how much easier it is when metals, such as aluminum cans, are turned directly into new aluminum cans, or when steel and glass containers are turned directly into new products. Plus, by reusing the existing resources, there would be a tremendous cost savings on energy and landfill space, as well as environmental impacts. This is why the triangle-shaped symbol is placed on products that can be recycled (or it can also denote the item comes from recycled materials). The symbol was based on the "Mobius Strip." ♺ In the 1850s, August Ferdinand Mobius, a Mathematician, came up with the idea for a science mathematical contest in Paris, France. His unique idea was to take a rectangle shaped piece of paper, make a simple single twist in it, and tape the two ends together. By doing so, he produced a two-dimensional surface with a single continuous edge. When flattened out, it takes on a triangle shape. (Mobius, n.d.) It was a perfect match to how nature works with a constant, continuous flow of nutrients through our biosphere. Ecologists/Environmentalists adopted it as the symbol of working with nature's processes. (Notes 23, page 57) As I would always remind my students, and as I'll remind my reader throughout this book, we live in a "closed system." What we do, or fail to do, with our resources, can and will have a lasting impact on our only home, for the good and bad. Remember the words of Chief Seattle "…What befall the Earth, befalls all the sons of the Earth…"

Here's a fitting instrumental song by Ennio Morricone's ♫ *The Good, The Bad, and The Ugly* ♫ (1967). It's so Boss!

23 | **THE MINERAL CYCLE, HUMAN IMPACTS AND RECYCLING**

"THE LAW OF CONSERVATION OF MASS"

IN = OUT

MELT IT, BURN IT, RECYCLE IT,

LANDFILL IT → IT'S STILL WITH US!

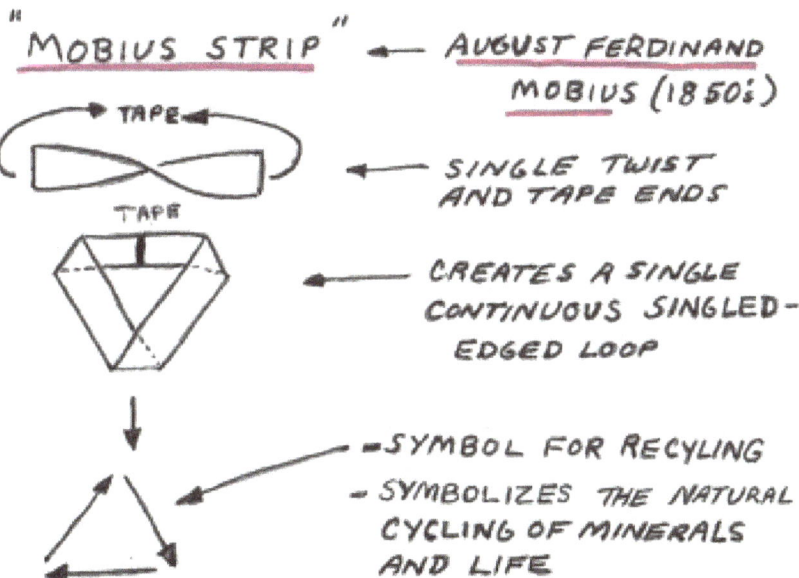

"MOBIUS STRIP" ← AUGUST FERDINAND MOBIUS (1850's)

TAPE

TAPE ← SINGLE TWIST AND TAPE ENDS

← CREATES A SINGLE CONTINUOUS SINGLED-EDGED LOOP

— = SYMBOL FOR RECYLING
— SYMBOLIZES THE NATURAL CYCLING OF MINERALS AND LIFE

REMEMBER "WE ARE ALL CONNECTED…"

THE CARBON CYCLE

B | **THE CARBON CYCLE** "THROUGH THE ECOSYSTEM"

24

* CLOSED SYSTEM

Ex.)

ELEMENTS THAT MAKE UP HUMANS

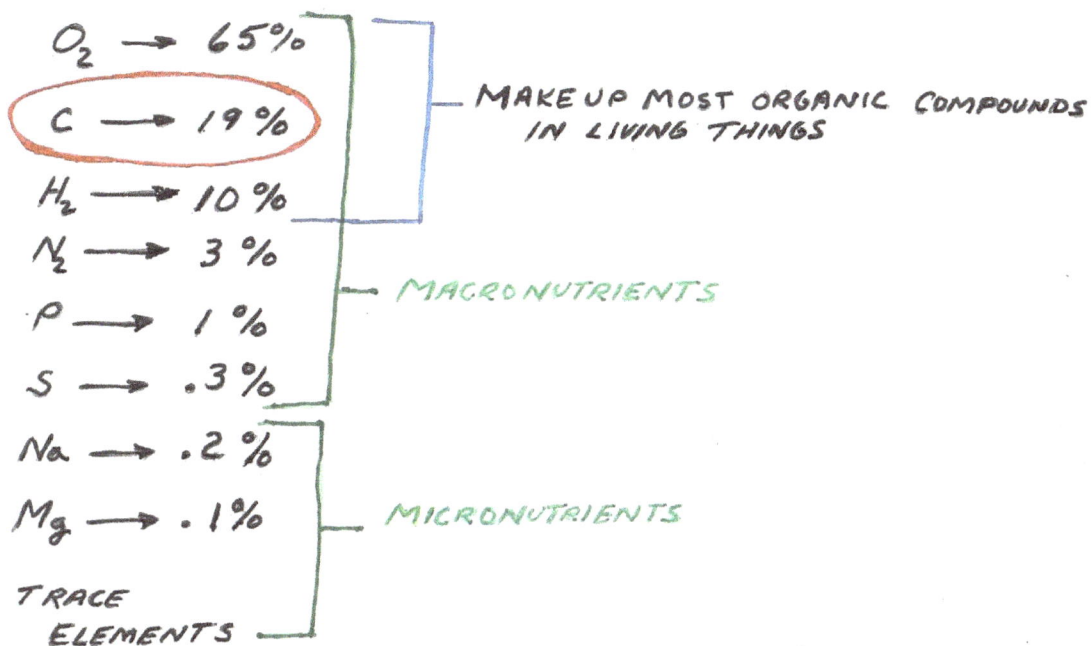

$O_2 \longrightarrow$ 65%

$C \longrightarrow$ 19%   ⎫ MAKE UP MOST ORGANIC COMPOUNDS
                              IN LIVING THINGS
$H_2 \longrightarrow$ 10% ⎭

$N_2 \longrightarrow$ 3%

$P \longrightarrow$ 1%   ⎫ MACRONUTRIENTS

$S \longrightarrow$ .3%

$Na \longrightarrow$ .2%

$Mg \longrightarrow$ .1%   MICRONUTRIENTS

TRACE
ELEMENTS

"C" $\longrightarrow$ MAKES UP $\longrightarrow$ LIVING THINGS
            ORGANIC COMPOUNDS

Hi

What would life be like on Earth without carbon? Carbon, along with varying amounts of Hydrogen and Oxygen, is the main element that makes up most of the molecules found in living things. (Notes 24, page 59; Notes 25, page 61) Carbon has this unique ability to bond with other elements to form inorganic and organic molecules and compounds of carbon. Carbon, because of its electron arrangement, loves to bond with other elements, and oxygen is one of its favorites. When one oxygen combines with one carbon, it creates carbon monoxide (CO), a very deadly gas that results from the incomplete combustion of carbon-based materials. (Notes 26, page 62) Carbon can also become carbon-dioxide (CO_2) with the addition of two oxygens. The carbon-dioxide (CO_2) molecule becomes the very gas that many organisms consume or release as the result of their biological processes. (Notes 26, page 62; Notes 27, page 65)

The carbon cycle begins as carbon-dioxide (CO_2) as a gas found in all three parts of our biosphere - land, water, and especially, the atmosphere. What's truly fantastic about carbon is its relationship with autotrophs in our biosphere. This relationship is called "Photosynthesis." These guys can take up the carbon-dioxide (CO_2) from their surroundings, and with the help of the sun's energy, can fix carbons into the production of organic-based molecules and compounds.

25 "C" LOVES TO BOND TO OXYGEN

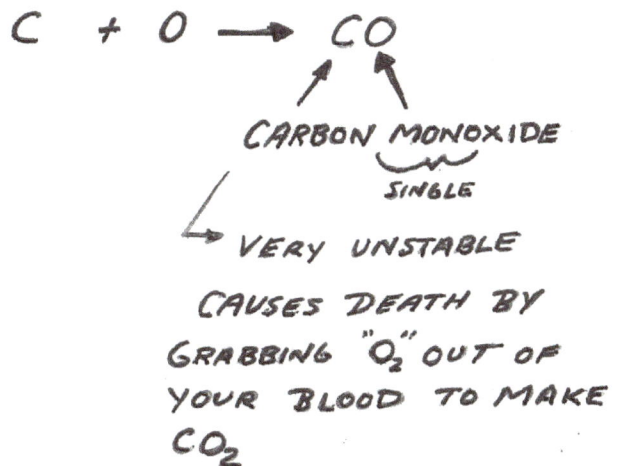

EX.) ATMOSPHERE
SOIL
WATER
} $C + O_2 \rightarrow CO_2$

CARBON DIOXIDE
2

$C + O \rightarrow CO$

CARBON MONOXIDE
SINGLE

VERY UNSTABLE

CAUSES DEATH BY GRABBING "O_2" OUT OF YOUR BLOOD TO MAKE CO_2

CARBON BONDING

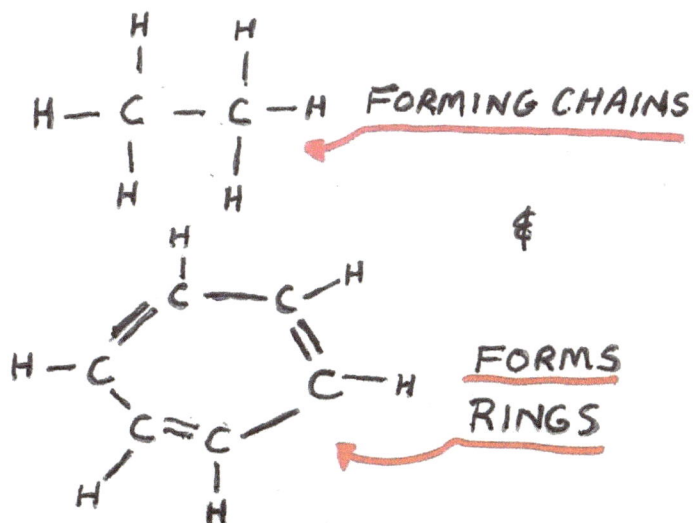

FORMING CHAINS

&

FORMS RINGS

CARBON CYCLE

$H_2O + CO_2 + E \rightarrow$

CO_2 CO_2 CO_2 CO_2

EX. — TREES
— GRASSES

GIVE OFF O_2 ☆ LOOK!

THE CARBON
IS FIXED INTO
THE TISSUES OF
THE PRODUCERS

GIVE OFF
O_2

RELEASED CO_2

3 ROUTES
"C"
CAN TAKE

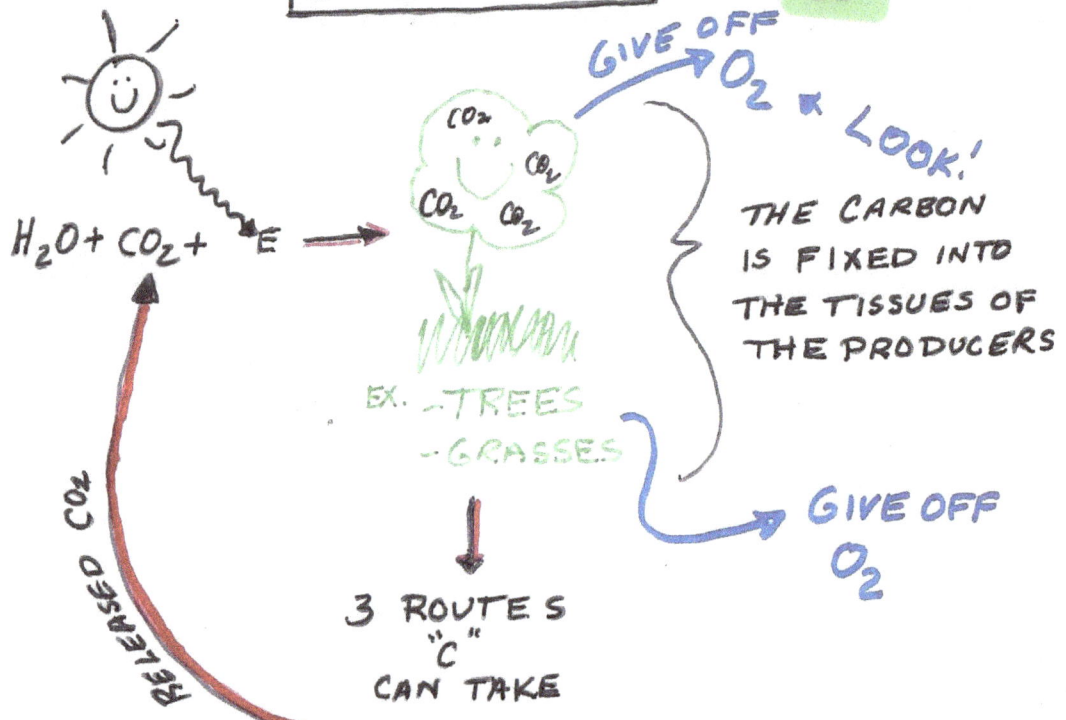

1.) RELEASED BY
FIRE   EX.) FOREST
OR CAMPFIRE

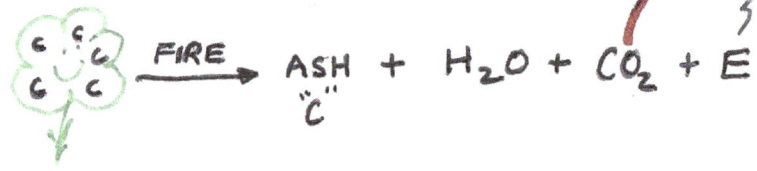

C C C C $\xrightarrow{\text{FIRE}}$ ASH + H_2O + CO_2 + E
"C"

LOST
AS HEAT
"E"

SOME "C" IS
RELEASED BACK
AS CO_2

2.) THE PRODUCER
DIES AND GOES
RIGHT TO THE
DECOMPOSER

SOME "C"
BECOMES
THEIR
TISSUES
YUM

C C C

YUM
YUM

$+ CO_2 + H_2O + E$

LOST
AS
HEAT

These carbon-based products eventually become the tissue components of the producers. Here is what the official reaction looks like:

$$6H_2O + 6CO_2 + ENERGY\ (E) \rightarrow C_6H_{12}O_6 + 6O_2$$

For simpler and more practical purposes, our photosynthesis reaction will look like this:

$$H_2O + CO_2 + E \rightarrow \text{❀} + O_2$$

↗
The tissues of all kinds of producers

Notice in the official reaction the coefficients, or numbers, next to carbon-dioxide (CO_2) water (H_2O) and oxygen (O_2), these numbers balance the photosynthesis reaction. Remember the "Law of Conservation of Mass?" – what goes in will equal what eventually comes out. Notice the products of photosynthesis - carbohydrates (energy molecules used to make plant tissues) plus the release of oxygen. "Oh," yes! "I love oxygen doesn't everyone?" This when most people would give me that "Toto, I don't think we're in Kansas anymore" look. I tell everyone if you are in doubt of your love for oxygen, to just put a paper bag over their heads for a few minutes. I would remind them of *"what's loves got to do with it."* In turn, my question to you is "where does the majority of the oxygen we take in come from?" Most would say the trees and/or the fields of grass. That's when I would "Gong" all of them (The Gone Show, 1976).

And then ask the question "What about winter." There're no leaves on the trees. The grasses are either frozen or covered up with snow. Again, I ask you to think about how is it we have enough oxygen during the winter? What does everyone hold their breathe all winter?" I don't know about everyone else, but I love my oxygen year-round. The answer, the majority of our oxygen is supplied to us by way of the oceans.

While the boreal and rain forests produce a lot of oxygen, the larger portions are supplied by the oceans. The oceans' aquatic plants, algae, and phytoplankton carry on the process of photosynthesis producing the bulk of our oxygen in full production year-round. The sun's energy is striking the Earth's oceanic areas somewhere around the world all the time. When the sun sets, the autotrophs stop their photosynthesis process. However, they don't stop their cellular processes that keep them alive until daybreak. At night, they rely on cellular respiration using the stores of carbohydrates/sugars manufactured during the day by photosynthesis. This is where many of my readers are probably having that look straight out of Shrek (2001) "Who is this Farquaad guy...?" Many people quickly reason, "if we are carrying on respiration right now and they're carrying on respiration at night, how do we get our oxygen that we love so much?" The answer, is not to worry because the oceans are so efficient at carrying on photosynthesis, they produce way more oxygen during the day than they could possibly consume at night. This leaves large reservoirs of oxygen in our atmosphere to accommodate all the consumers who use it for cellular reparation, and this includes humans. In fact, the oceans are known as a "Sink" for carbon dioxide (CO_2). The analogy behind the term "Sink" comes from the conventional kitchen sink, with a drain, one can find in most homes. The water goes down into, and is taken up by, the sinks drain. The water going down a kitchen drain is comparable to the oceans' water acting like a sink drain by absorbing large amounts of carbon dioxide (CO_2) out of our atmosphere. Luckily, for most organisms of the biosphere, the autotrophs release large amounts of oxygen back into the atmosphere to be circulated around the globe.

As a side note, with the oceans suppling the bulk of our oxygen, it would seem to me that if there is one portion of our biosphere so critical to our well-being, we would make sure to take extremely good care of it. And we should. But we don't. The human impact on the oceans is well documented. They are being overfished, polluted with oil slicks, as well as floating islands of trash made up mostly of discarded plastics.

3.) CAN BE TAKEN UP BY A CONSUMER

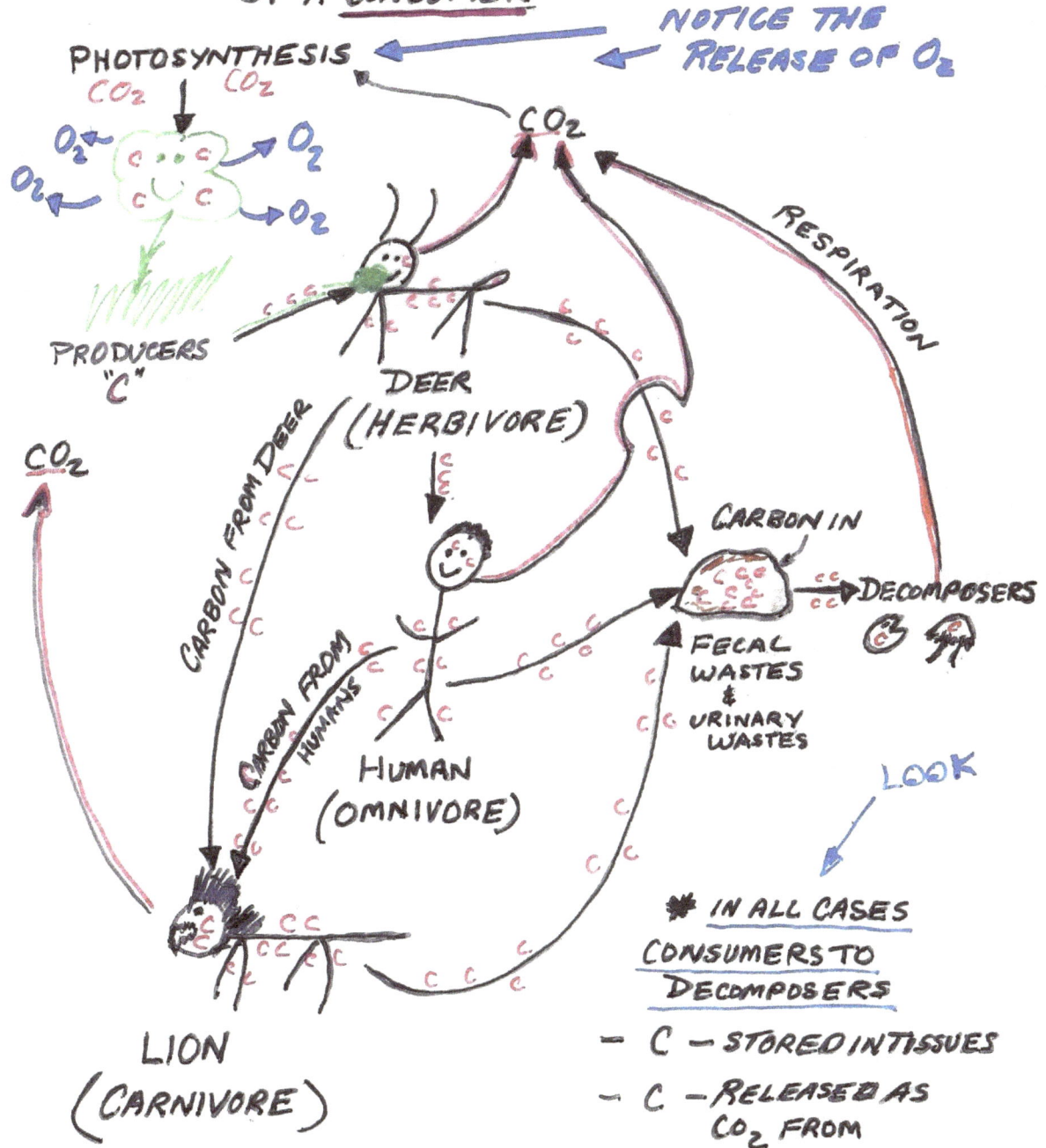

NOTICE THE RELEASE OF O_2

PHOTOSYNTHESIS

CO_2 CO_2

O_2 O_2 O_2 O_2

CO_2

PRODUCERS "C"

RESPIRATION

DEER (HERBIVORE)

CARBON FROM DEER

CO_2

CARBON FROM HUMANS

HUMAN (OMNIVORE)

CARBON IN

FECAL WASTES & URINARY WASTES

DECOMPOSERS

LOOK

LION (CARNIVORE)

* IN ALL CASES CONSUMERS TO DECOMPOSERS

- C — STORED IN TISSUES
- C — RELEASED AS CO_2 FROM RESPIRATION
- C — IS PASSED AS URINARY & FECAL WASTES

That's just a few of the problems facing our oceans. (Parker, 2018, June) If everyone realized how interconnected all living things are on our planet, this would be a wakeup call.

All This oxygen stuff put another song in my head. Let's face it everyone loves oxygen, but everyone needs to be careful not to get too much because it could make a person feel high, or light headed; but remember, not enough and one could die.

Check out the song ♫ *Love is like oxygen* ♫ by Sweet (1978). Most people tell me have a song for everything!" I would reply "I do, I do, I must, I must." It's a fun way to reinforce a very important topic.

The Carbon Cycle, as I mentioned earlier, starts in the atmosphere and is taken up by the producers. Make sure to look at my notes 27, page 65; Notes 28, page 67; and Notes 29, page 69. Wow! If one is a visual learner, get ready for an informational ride of one's life. At first my arrows may seem, "♫ *Haltered Sheltered* ♫" (if one will excuse the song), but there are repeating patterns to carbons' ability to flow in and out of the cycle. All three pages of my notes overlap as intended. Nonetheless, I still like the carbon cycle overview of my notes the best. (Notes 28, page 67)

As I discussed earlier, carbon is fixed into the tissues of producers by the photosynthesis process. Only the autotrophs can do this remarkable process of absorbing and converting of inorganic materials, along with the sun's energy, to create the organic, carbon-based tissues of all living things on our planet. All life forms are dependent on the autotrophs and/or producers. Consequently, all heterotrophs/consumers receive their nutritional and energy requirements directly or indirectly from producers. Don't forget, these are the same critters (like us) that love the autotrophs because of their ability to release oxygen. If one follows the flow of carbon after it has been fixed into the tissues of producers, its return back to the environment has a common theme.

From producers to decomposers, their goal in life is to eat and reproduce. To do so, all life forms must nourish themselves on the nutrient, carbon-rich, tissues of other organisms. As consumers perform their cellular processes, they break apart the carbon structures to be incorporated into their carbon-based tissues for growth and reproduction. While some of the carbon-based nutrients consumed by organisms are directly used for their development, some are released back into the environment as wastes or the biproducts of their metabolism. When

THE CARBON CYCLE OVERVIEW

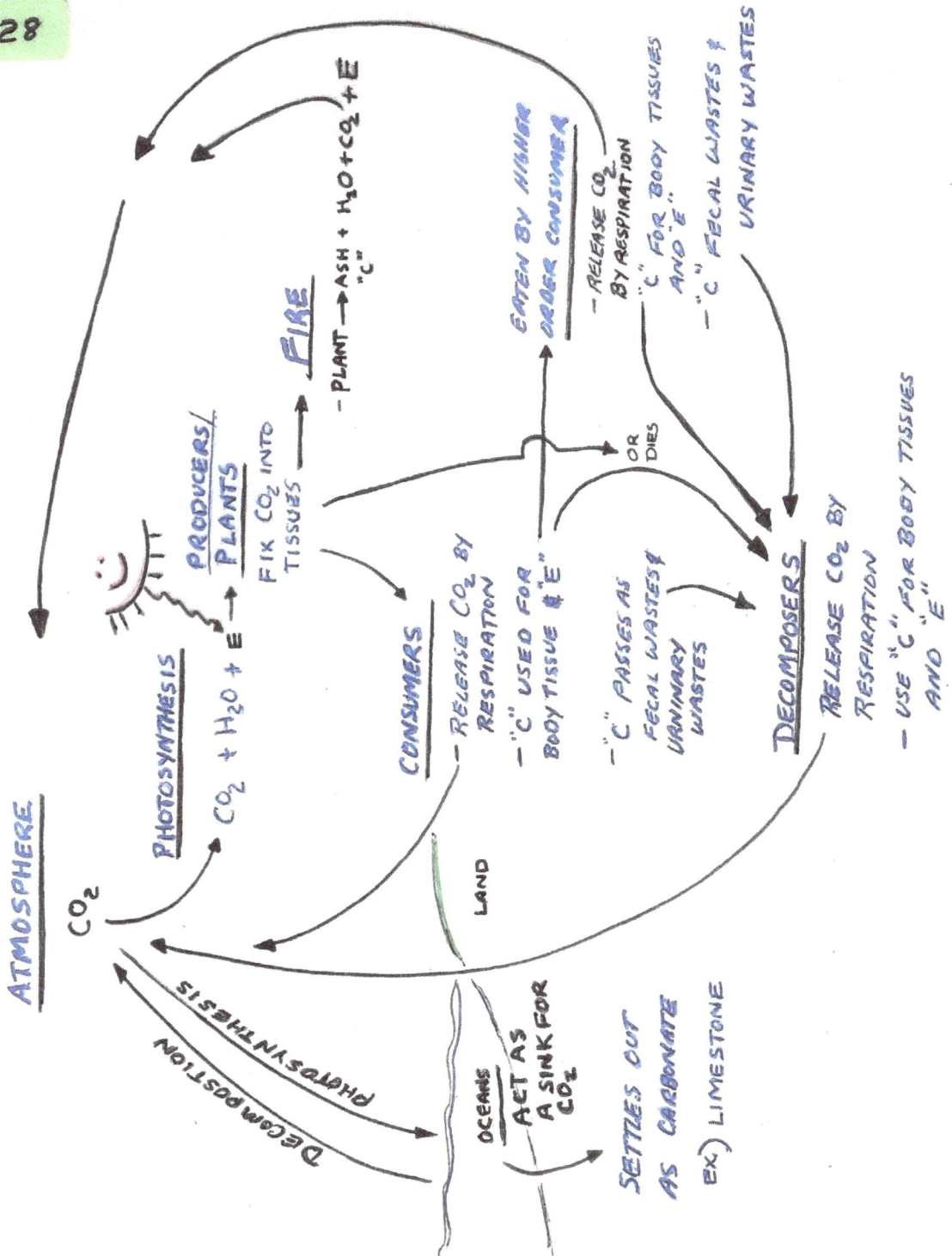

ATMOSPHERE

CO_2

PHOTOSYNTHESIS

$CO_2 + H_2O + E$

PRODUCERS / PLANTS

FIX CO_2 INTO TISSUES

FIRE

- PLANT $\rightarrow$ ASH + $H_2O + CO_2 + E$
  "C"

EATEN BY HIGHER ORDER CONSUMER

- RELEASE CO_2 BY RESPIRATION
- "C" FOR BODY TISSUES AND "E"
- "C" FECAL WASTES & URINARY WASTES

OR DIES

CONSUMERS

- RELEASE CO_2 BY RESPIRATION
- "C" USED FOR BODY TISSUE & "E"
- "C" PASSES AS FECAL WASTES & URINARY WASTES

DECOMPOSERS

- RELEASE CO_2 BY RESPIRATION
- USE "C" FOR BODY TISSUES AND "E"

LAND

PHOTOSYNTHESIS

DECOMPOSITION

OCEANS

ACT AS A SINK FOR CO_2

SETTLES OUT AS CARBONATE

EX) LIMESTONE

organisms carry on their cellular functions, such as respiration, they release some of the consumed carbon back into the environment as carbon dioxide (CO_2). I also must point out that the amounts of carbon compounds processed, utilized, and released are different for each individual species. This results in carbon-based metabolic wastes of non-used or indigestible food materials that need to be expelled from their bodies as urinary and fecal wastes. In turn, these wastes become dinner on the plate for decomposers. In the end, even if a consumer outright dies, or is killed, like a dead skunk in the middle of the road (even in the absence of being eaten by a higher order consumer), the decomposers will eventually return a consumer's dead body of carbon and its nutrients back to the environment for reuse. Uh-ho, I just thought of another song to reinforce this concept. Unusual perhaps, go check it out ♫ *Dead Skunk in the Middle of the Road* ♫ by Loudon Wainwright III (1972).

The carbon cycle, like the mineral cycle, can be carried out as a short-term or long-term cycle. (Notes 29, page 69) A good example of a short-term cycling of carbon would be eating one's lunch. Right in one's mouth, even as one takes their first bite, there are metabolic processes breaking the bonds of carbon compounds found in the form of carbohydrates in the food. Within minutes to a few hours, the carbohydrates are quickly absorbed and moved to the cellular level for the respiration process. In turn, one's cells begin to release some of that carbon, back into the environment as carbon dioxide (CO_2). The release of carbon dioxide (CO_2) occurs through one's skin and lungs. As I described before, the same would hold true for the rest of one's lunch. The food material will move through the digestive track where the needed carbon-based nutrients will be absorbed and the unwanted carbon-based materials will be expelled as urinary and/or fecal wastes or in some cases, flat out regurgitated. As one would now be able to guess, the decomposers will finish the breakdown and final release of the carbon back into the biosphere for reuse. As a side note I'll cover the cellular respiration process in more detail later when I discuss energy flow in the ecosystem.

"C" CYCLE CAN BE... 29

SHORT TERM

    — VERY FAST CYCLE
        EX.) TAKEN IN AS CO_2 DURING
            PHOTOSYNTHESIS AND RELEASED
            DURING RESPIRATION

LONG TERM

    — VERY SLOW OR HELD IN STORAGE
        EX.) FIXED IN PLANTS
           EX.) LONG LIVING TREES
              REDWOODS → SEQUOIA
      OR
      IN LIMESTONE
      OR
      IN FOSSIL FUELS

Carbon can also be held in a long-term cycle. This time frame is all relative to the many individual organisms and their lifespans. For example, elephants can build up and store carbon in their tissues during a full lifespan, somewhere between sixty and seventy years, (Vitale, n.d.), while redwood trees can store their carbon in their tissues for thousands of years (Berg, 2017). An even longer time frame involves carbon being held in the tissues of dead plants and animals trapped underground only to be decomposed under extreme pressures, temperatures, and biological processes – sixty-five million years in the making. Also known as fossil fuels, coal, oil, and natural gas. (Notes 30, page 71)

This is the process of fossil fuel development I started to describe earlier with the mineral cycle. I will continue with more detail on this topic in some of the other sections because of its uses and the significance on our ecology. Remember, "…The wind that gave our grandfather his first breath also receives his last…" (Chief Seattle, 1854)

There were many story songs written and played about the coal miners; ♫ *Big John* ♫ by Jimmy Dean (1961) was one of those. Can you "dig it!"

FOSSIL FUELS

- REMAINS OF ORGANISMS THAT WERE TRAPPED, DECOMPOSED, AND COMPRESSED UNDERGROUND CREATING FUELS USED TODAY

"E" FROM THE SUN IS TRAPPED IN PLANTS AS CARBON BONDS

EX.) SUGARS, TISSUES

THE REMAINS OF ANCIENT CRITTERS COLLECTED AND TRAPPED, ISOLATED BY THE EARTH'S LAYERS

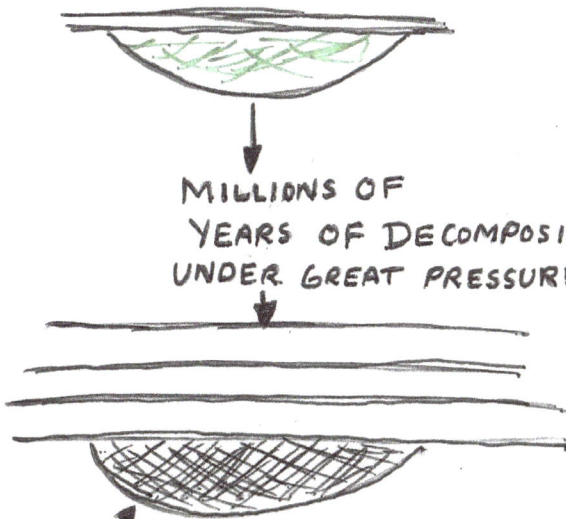

MILLIONS OF YEARS OF DECOMPOSITION UNDER GREAT PRESSURE

CREATING CARBON COMPOUNDS RICH IN "E"

OIL, COAL, GAS

THE FOSSIL FUELS WE USE TODAY!

The Carbon Cycle: The Human Impacts

By understanding the natural cycling of nutrients through the biosphere, it should be a whole lot easier to understand the impacts on our planet of using fossil fuels in such staggering amounts over the last 150-200 years.

The extraction, processing, and burning of fossil fuels releases a variety of organic-based gases. While there are a few gases involved in global warming, it's been scientifically proven that carbon dioxide (CO_2) along with methane (CH_4) are the two main gases causing the problem. (Arms, 2008) To simplify the scope of global warming and its impacts on our biosphere, my focus will be on the main global greenhouse gas - carbon dioxide (CO_2). My goal here is to provide a streamlined, practical understanding of global warming (also known as the greenhouse effect) as it pertains to man's impact on the use of fossil fuels and the carbon cycle. However, before I can explain the impacts of carbon dioxide (CO_2) on our biosphere, one needs to know a little background information on how we, as humans, developed a world economy built on the use of fossil fuels that ultimately led to the ecological problems we face today.

The mining and burning of coal led to advances in the new industrial age in the latter part of the 1800's by supplying the high demand for energy and carbon requirements for steel production. With continued advances during this time period, manufacturing quickly turned to oil, another form of fossil fuels, that was needed to lubricate the new machinery that was increasing the efficiencies and production of goods. Eventually, electricity would be seen as an energy resource. Not only to light up homes and factories, but to further the needs of the growing industrial age. In order to produce large volumes of electricity, generators would need to be turned. Again, industry turned to coal. Coal stoked the fires of the boilers to heat water into high pressure steam that would be released to turn the propeller-like blades that were housed in controlled units called turbines. In turn, the turbines would turn the shafts of the generators to produce high volumes of electrical current. This process will be detailed in the section on energy flow.

As the number of workers continued to grow cities began to expand. Plus, industry was expanding with an ever-increasing demand of energy and raw materials for the production of finished goods. With greater industrial output and opening of new markets, larger and more efficient transportation systems were required. More railroads were being built and steam powered shipping took control of the oceans, all of which required larger amounts of energy provided by fossil fuels. Then came the production of automobiles and road systems to drive them on. The resulting commercial and world economics that were developed with fossil fuels made our lives easier. As populations grew and industrialized nations developed, the demand for more electrical power plants, airplanes, and automobiles increased the demand and use of fossil fuels. And with the increased demand and use of fossil fuels came air pollution. We started to spew out so many chemicals and particulates into our atmosphere that we were causing noticeable changes in the Earth's atmosphere and climate. (Brenarde, 1973)

Environmentalists started to research the effects of the chemicals being released from the extraction, production, and burning of fossil fuels. Researchers knew that when burning any organic compound to release its heat energy, it also released carbon in the form of carbon dioxide along with water. Remember, the Law of Conservation of Mass? Under controlled conditions, scientists proved carbon dioxide (CO_2) has the ability to hold heat energy. They also knew large volumes were being released into the atmosphere and their concentrations were building up. In fact, with the large volume of carbon dioxide being released, the autotrophs, by way of their photosynthesis process, could never offset the volume being released. To prove this, molecular measurements of carbon dioxide in the atmosphere were documented, and the concentrations were increasing at a noticeable rate.

Here's how the process of global warming works. Make sure to look at my Notes 31, page 75; Notes 32, page 76; and Notes 33, page 78. Our atmosphere has three main functions. The first is to filter out dangerous solar radiations. Second, it connects the land and aquatic ecosystems. Third, it acts as a big regulator by holding in heat energy as our planet makes its daily rotation. Without this function, our planet would only know two extremes – super hot and super cold, not a real good scenario for life to exist. To act as this protector, connector, and insulator, our atmosphere is made up of varying amounts of gases. The atmosphere contains about 78% nitrogen and 21% oxygen. Yet, that last one percent of other gases, mainly carbon dioxide and methane gases, became significant in the global warming problem. (Andrews, 1972)

As I discussed earlier, it has been proven that under controlled conditions, these gases are very good at holding heat energy. Consequently, with all of the released carbon dioxide into our atmosphere, Earth's average temperature has now started to increase. Remember we live in a closed system. The main source of all this carbon dioxide buildup is the net result of using and burning fossil fuels that are carbon-based products which come directly or indirectly from dead plants and animals. As discussed earlier (Notes 30, page 71), these are organisms from another time in the Earth's history. Millions of years ago, these carbon compounds of fossil fuels flourished as critters in a much different environment with a totally different atmosphere and climate than the one we know today. It's been well established with radioactive dating and physical evidence that there have been numerous mass extinctions in the Earth's history. For instance, cataclysmic events such as asteroids, volcanic eruptions, and plate tectonics, just to name a few, were responsible for significant amounts of life lost in our planet's history. (Velikovsky, 1950)

31

HOW GLOBAL WARMING / GREEN HOUSE EFFECT WORKS

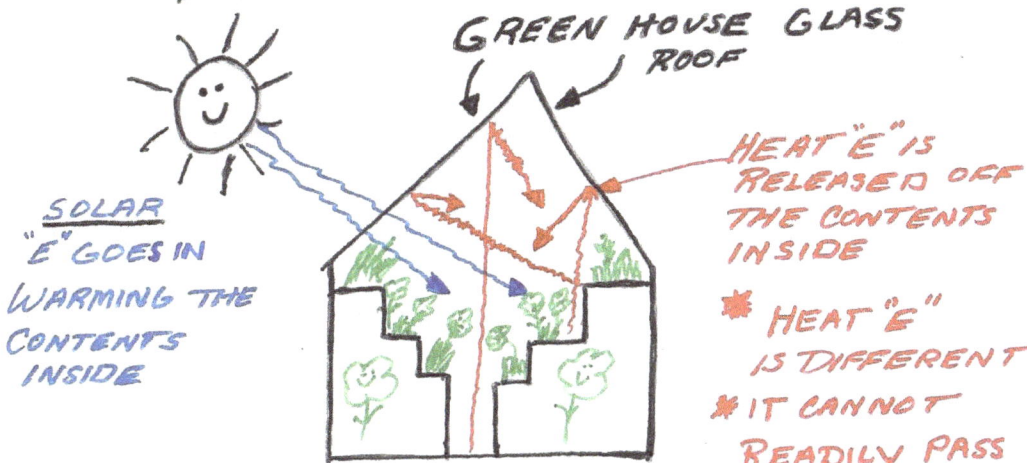

GREEN HOUSE GLASS ROOF

SOLAR "E" GOES IN WARMING THE CONTENTS INSIDE

HEAT "E" IS RELEASED OFF THE CONTENTS INSIDE

✳ HEAT "E" IS DIFFERENT
✳ IT CANNOT READILY PASS THROUGH THE GLASS - BEING TRAPPED

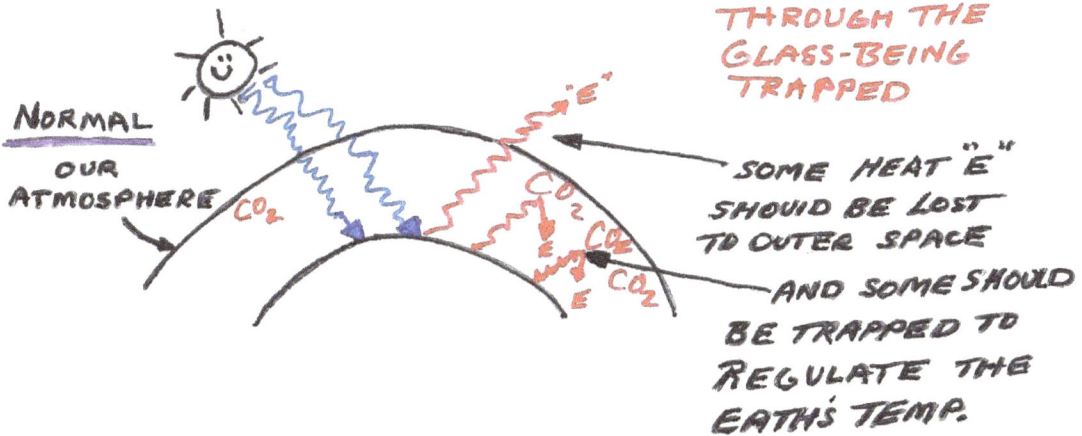

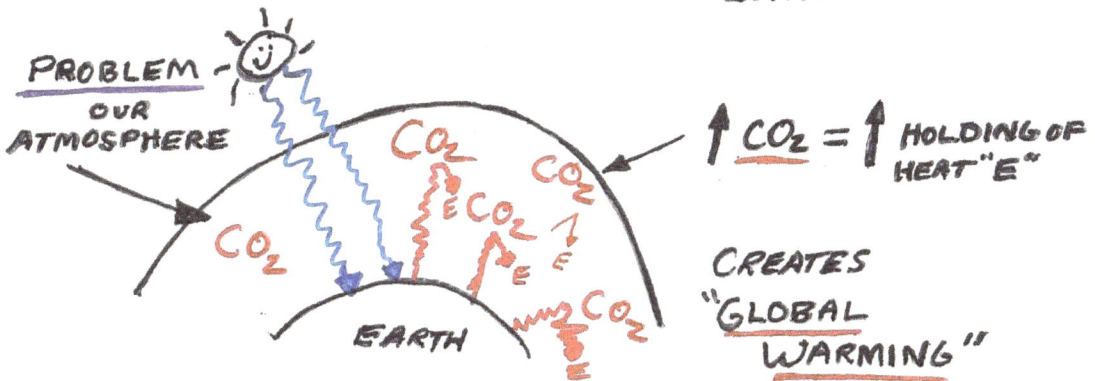

NORMAL
OUR ATMOSPHERE
CO_2
"E"
CO_2
CO_2

SOME HEAT "E" SHOULD BE LOST TO OUTER SPACE

AND SOME SHOULD BE TRAPPED TO REGULATE THE EATH'S TEMP.

PROBLEM
OUR ATMOSPHERE
CO_2
CO_2
CO_2
CO_2
CO_2
CO_2

EARTH

↑ CO_2 = ↑ HOLDING OF HEAT "E"

CREATES "GLOBAL WARMING"

32 WHY THE "C" CYCLE IS IMPORTANT
"CLOSED SYSTEM"

* RELEASE OF TOO MUCH CO_2 IS CAUSING OUR
BIOSPHERE → ATMOSPHERE / CLIMATE TO
CHANGE

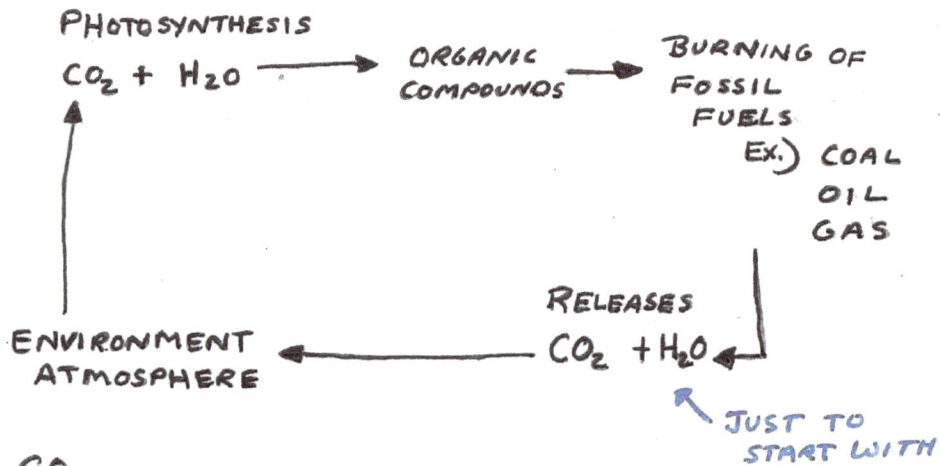

PHOTOSYNTHESIS
$CO_2 + H_2O$ ⟶ ORGANIC   ⟶ BURNING OF
COMPOUNDS   FOSSIL
FUELS
Ex.) COAL
OIL
GAS

ENVIRONMENT ⟵ RELEASES
ATMOSPHERE $CO_2 + H_2O$ ⟵

JUST TO
START WITH

CO_2 IS KNOWN TO HOLD HEAT
IN OUR ATMOSPHERE

↑CO_2 LEVELS ARE CAUSING
"GLOBAL WARMING"

⟶ WE ARE ALREADY SEEING
THE CONSEQUENCES OF "GLOBAL
WARMING"

DISRUPTS: ECOSYSTEM
STRUCTURE

LOSS OF FUNCTION
ALLOWING LIFE TO EXIST

These cataclysmic events, resulted in changes in the Earth's atmosphere, climate, as well as its landscape, burying these carbon-based tissues underground, placing them in a very long-term storage. Now, with large operations of coal mining, oil field drilling, and production of natural gas, all of which are being burned at an ever-increasing rate, these fossil fuels are releasing their carbons back into today's atmosphere. By burning these enormous amounts of fossil fuels, we are releasing large volumes of global greenhouse gases. In turn, the increased carbon dioxide concentrations are holding more of the Earth's heat energy that results from the sun's solar energy striking the Earth.

As I mentioned earlier, the atmosphere is normally supposed to hold a certain amount of the Earth's heat energy, keeping it from outright freezing everything to death every night. Nonetheless, the atmosphere is supposed to lose a portion of that heat energy on a daily basis keeping the Earth's temperature regulated. Again, with increased concentrations of carbon dioxide, that heat energy is being held back from escaping from the Earth's atmosphere. As a result, the atmosphere and the average global temperature starts to rise. This is global warming.

A great analogy was presented to me years ago that explains what global warming is comparable to. Most everyone knows what it's like to hop into a car that has been sitting in the hot summer sun for hours. The same premise is used to intentionally heat large glass buildings and grow plants during the cooler and cold weather months. (Notes 31, page 75) The solar rays come through the glass heating up the materials inside the greenhouse. In this case scenario, the car becomes the greenhouse. Inside the car, the solar light energy strikes and heats up all the upholstery, seats, dashboard, etc. These heated materials release the heat energy back into the car's atmosphere but are trapped by the closed glass windows, increasing the inside temperature of the car. So, as one jumps into their car on a hot summer day, down go the windows of the car to let out the heat. Now, think of the Earth's atmosphere like that of a car sitting in the sun. Only now one starts to slowly roll the windows up. Consequently, the heat energy has less of an exit. The same would be comparable to the carbon dioxide concentration acting like the glass holding in the heat, and the more carbon dioxide we release, the more we roll up the windows of our atmosphere. (Eunson, 1990, Jan. 1)

PREDICTED OUTCOMES RESULTING FROM GLOBAL WARMING

MELTING ICE FROM WORLD GLACIERS AND POLAR REGIONS

↓

HELP I CAN'T SWIM

INCREASED FLOODING OF COASTAL SHORELINES

↓

LOSS OF WILDLIFE AND DISPLACEMENT OF HUMAN POPULATIONS

CHANGES IN WEATHER PATTERNS

— LESS SNOW AND INCREASED MELTING OF SNOW AND GLACIAL ICE IN MOUNTAINOUS REGIONS

I'M THIRSTY

↳ LESS FRESH WATER REPLENISHMENT

— DRIER REGIONS W/ INCREASED NUMBERS AND SIZES OF FOREST FIRES

— INCREASED HURRICANES W/ GREATER INTENSITY W/ INCREASED OCEAN TEMPS.

Since the building up of carbon dioxide concentrations in our atmosphere, the Earth's average global temperature has gone up resulting in many negative and/or potential catastrophic impacts on our biosphere. For example, one only has to look at pictures taken of glacial ice sheets from the early 1900s and compare them to recent pictures taken from the exact same spots. They are retreating at an unnatural rate. This is a net result of ice melts coinciding with increased global temperatures that are changing our planet's climate. These climate changes are having a direct impact on our weather systems. Less snow is being deposited with more glacial melting in mountainous and polar regions of the world. With less snow deposits in mountainous regions, there will be significant drops in melt-offs that regularly replenish many freshwater systems. This means less fresh water for large cities and their residents. Conversely, the melting waters of the glacial and arctic ice sheets will have the opposite effects on ocean levels. Water that was once tied up in these large volumes of ice will cause ocean and sea levels to rise, creating higher tidal actions and a significant loss of shoreline. Subsequently, large populations of plants, animals, as well as many other micro- and macroscopic organisms, will be deeply impacted if not outright lost because of their habitat destruction. Not to mention the massive relocation of large amounts of people who now live in many large coastal cities. (Notes 33, page 78)

In addition, the increased global temperatures are dramatically impacting our planet's weather patterns. These changes are causing some areas of our planet to become even drier, increasing the number and scope of massive forest fires. One only has to look at the record number and size of fires that have occurred across the western portions of the United States over the last couple of years (Westerling, 2006 Aug. 18). Also predicted and coming to pass is the warming of the ocean's average temperature. There are many species of plants, animals, and other marine life, as well as whole ecosystems, that are very temperature sensitive and being affected, such as the great barrier reef off the northwest coast of Australia (Meyer, 2018, Apr.18). This is only one of many examples being documented about increased ocean temperatures impacting wildlife and their ecosystems. (Notes 34, page 80)

Another concern coming to pass with the increased ocean temperatures is the global hurricane season that is beginning earlier, lasting longer, and having ever-increasing intensity. As I write these pages, the United Nations has just released its updated report on the global warming situation. Its potential effects on life on Earth will be even worse if this problem is not taken seriously (Davenport, 2018 Oct.7).

Wow! This is just a sampling of the impacts that the carbon cycle is having, and will continue to have, on our biosphere. And, I didn't even touch on its effects on algae blooms, oil spills, or even the implications for crop productions and food supplies.

Remember "…The wind also gives our children the spirit of life…" (Chief Seattle 1854),

Another song just came to mind. I was thinking about clouds, and what's love got to do with it?" Judy Collins did, listen to her song, ♫*Both Sides Now* ♫, (1967). Be there or be square. Check it out!

THE NITROGEN CYCLE

C. NITROGEN (N) CYCLE

3 MAIN "N" STORES IN NATURE

1.) ATMOSPHERE

- CONTAINS 78% N_2 GAS
- INORGANIC ELEMENT

* MUCH OF THIS IS UNUSEABLE FOR ORGANISMS IN THIS FORM

"UNLIKE CO_2"

2.) INORGANIC "COMPOUNDS"

NITROGEN LOVES TO BOND W/ OXYGEN AND HYDROGEN WHEN "E" IS AVAILABLE

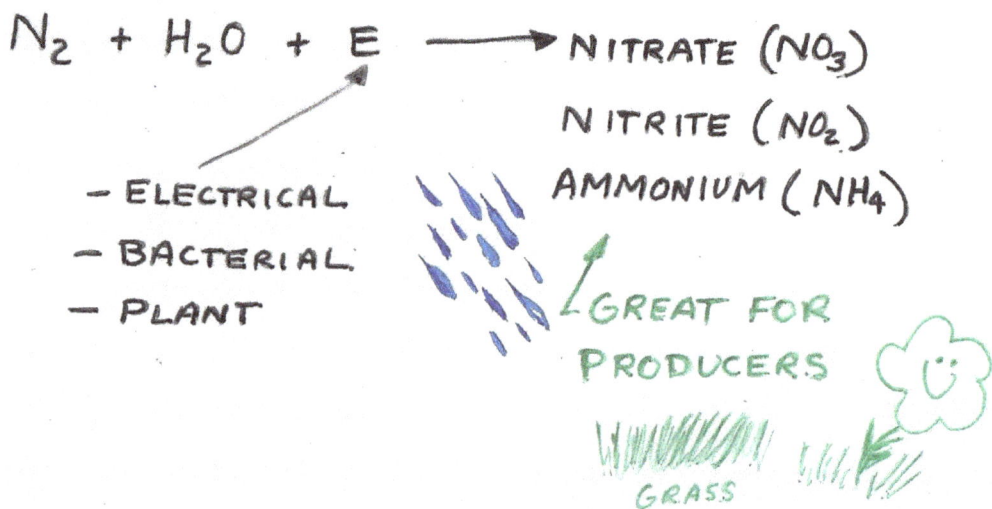

$$N_2 + H_2O + E \longrightarrow$$

- ELECTRICAL
- BACTERIAL
- PLANT

NITRATE (NO_3)

NITRITE (NO_2)

AMMONIUM (NH_4)

GREAT FOR PRODUCERS

GRASS

3.) <u>ORGANIC COMPOUNDS</u>

EX.) <u>PROTEINS</u> → FOUND IN TISSUES OF LIVING AND DEAD ORGANISMS

<u>PRODUCTS OF METABOLISM</u>

- FECES
- URINE → TURNS INTO → UREA

UREA IS UNUSED NITROGEN RELEASED FROM METABOLISMS OF ORGANISMS

EX.) OUR "KIDNEYS" REMOVE THIS FROM OUR BLOOD

| "N" IN THE ECOSYSTEM |

- "N" IS A MACRONUTRIENT FOR PRODUCERS MOSTLY PLANTS → FOR PROTEINS AND TISSUES

- CONSUMERS NEED IT FOR THEIR PROTEINS / TISSUES

The nitrogen cycle is another nutrient cycle that plays a significant role in the survival of all life in the biosphere. It's similar to the mineral and carbon cycles as it passes through the environment, from producers and/or autotrophs to consumers and/or heterotrophs to decomposers. However, it also differs (making it unique) in the ways it has to be converted into different forms for its eventual uptake and return to the biosphere. The biosphere has three stores (i.e. natural storages) where nitrogen can be found. To further my explanation of a nitrogen store concept one only has to be reminded why we call the grocery store, the grocery store, "Because they store or hold food for people to come and buy." Another follow up question to think about is "what would I buy at a hardware store?" Ok, hardware – good. (Notes 36, page 82; Notes 37, page 83)

The nitrogen cycle, like the carbon cycle, begins in the atmosphere, the first major store of nitrogen in our environment. As previously mentioned, 78% of our atmosphere is made up of nitrogen (N_2) gas. However, much of this nitrogen gas is in the form of N_2, making it unusable for many producers and/or autotrophs to take up directly, and this would hold true for almost all the consumers in the biosphere as well. It is in the second natural store of Nitrogen, inorganic nitrogen-based compounds, that can be readily used by most organisms. This is where atmospheric nitrogen (N_2) gas is fixed into inorganic molecules of nitrogen, for example nitrite (NO_2), nitrate (NO_3), and ammonium (NH_4). Just like carbon dioxide (CO_2), the nitrogen (N_2) gas likes to bond with oxygen (O_2) giving it different chemical characteristics from its original N_2 form. However, this is where nitrogen (N_2) gas needs a little help to bond with oxygen (O_2). In order for this bonding to occur there has to be an input of energy. This source of energy can be obtained naturally from lightning, or electrical current, bacteria and plants. When provided with the necessary energy these nitrogen (N_2) gases, along with water, produce inorganic forms of nitrogen that producers and other autotrophs love to chow down. When these inorganic forms of nitrogen-based compounds are taken up by the producers and/or autotrophs, they become fixed into their tissues. Once these nitrogen molecules are incorporated into the tissues of living things it creates the third storage of nitrogen - organic-based nitrogen compounds. (Notes 37, page 83)

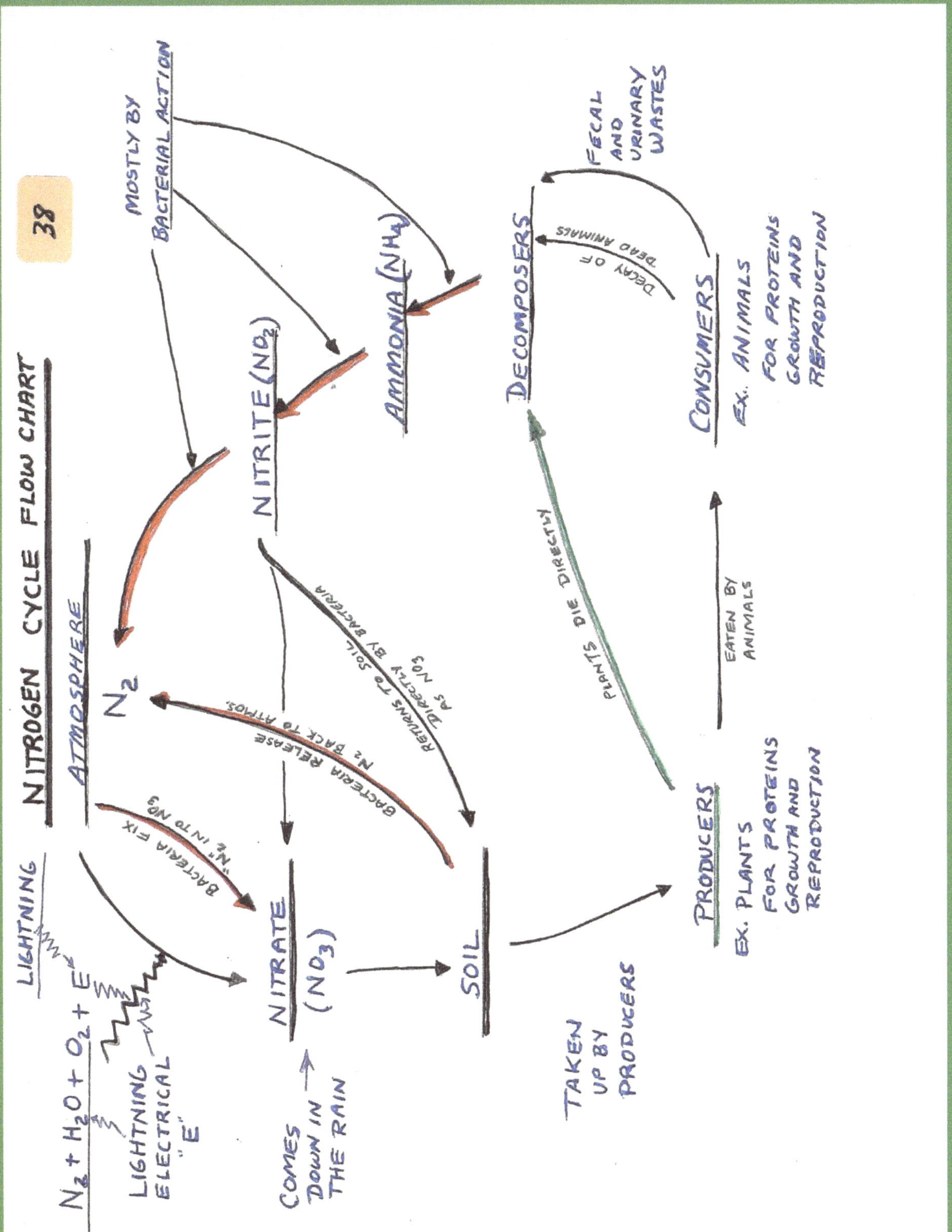

NITROGEN CYCLE FLOW CHART

38

MOSTLY BY BACTERIAL ACTION

AMMONIA (NH₄)

NITRITE (NO₂)

ATMOSPHERE

N₂

LIGHTNING

N₂ + H₂O + O₂ + E⁻

LIGHTNING
ELECTRICAL
"E"

BACTERIA FIX
"N" INTO NO₃

BACTERIA RELEASE
N₂ BACK TO ATMOS.

RETURNED TO SOIL
AS NO₃ BY BACTERIA

COMES
DOWN IN →
THE RAIN

NITRATE
(NO₃)

SOIL

TAKEN
UP BY
PRODUCERS

PRODUCERS

EX. PLANTS
FOR PROTEINS
GROWTH AND
REPRODUCTION

PLANTS DIE DIRECTLY

EATEN BY
ANIMALS

DECOMPOSERS

DECAY OF
DEAD ANIMALS

FECAL
AND
URINARY
WASTES

CONSUMERS

EX. ANIMALS
FOR PROTEINS
GROWTH AND
REPRODUCTION

These organic forms are made as they become the components of cellular structures for growth, repair, and reproduction. For instance, these organic-based compounds are eventually used as the building blocks for the DNA code, the instructional code for all lifeforms on our planet. The genetic code of life is made up of four nitrogen base molecules that code for everything that cells need to survive and reproduce. This includes coding for enzymatic catalysts, protein production, and the overall cellular structures that make up individual cells or eventually whole-body tissues. Consequently, it's an essential part of the nutrient cycling process. I believe most people have heard the phrase "we need a good rain to make the garden grow," well there is truth to that, and the nitrogen cycle plays a big role. Here's how rain water picks up nitrates and begins a crucial step in the nitrogen cycle. Remember to look at the nitrogen flow chart (Notes 38, page 85) as I discuss the details of the cycle.

Atmosphere lightning (a natural electrical energy from the charged particles) fixes the nitrogen (N_2) gas with water (H_2O) and oxygen (O_2) into nitrates (NO_3). These newly formed inorganic nitrates come down with the rain, making their way into the soil. Again, producers use large amounts of nitrates (a macronutrient) as they grow and reproduce. (Notes 17, page 43) As consumers feed on producers to meet their nutritional and energy needs, some of the nitrogen-based products are absorbed and utilized to become parts of their cells and tissues. In other words, if one looks closely at the nitrogen cycle flow chart, unlike carbon dioxide (CO_2), nitrogen- based compounds cannot be released directly back into the atmosphere through cellular reparation processes. Once the nitrogen-based materials become the components of another individual, they will be held there until they are either replaced as part of an individual's growth and repair or until that individual outright dies and/or is consumed by a higher-level heterotroph. In either case, the cells and tissues shed, or the remains of dead bodies will be broken down by the decomposers which will eventually release the nitrogen-based compounds back into the ecosystem for reuse. As discussed earlier in nutrient cycling, the undigested and/or unabsorbed nitrogen-based wastes that are not utilized (or the wastes released from cellular, metabolic processes) are eventually excreted as urinary and/or fecal wastes. Again, that's food on the plate for decomposers, the real heroes of the nitrogen cycle (especially the bacteria). These guys have the right stuff, or biological processes, to break down waste products and tissues of other organisms, releasing their nitrogen-based compounds back into the environment for future use. The return of nitrates (NO_3) back to the soil begins with a series of decomposition reactions, essentially reversing the formation processes described earlier. (Notes 38, page 85) The first stage of biological action is to convert the organic wastes to ammonium (NH_4) which is quickly converted over to nitrites (NO_2). With further processing of bacterial action, the nitrites are quickly turned into nitrates (NO_3) that are available for reuse by other critters, mainly producers and other autotrophs.

Plus, there is an added reaction happening at the same time. As bacterial actions are converting the nitrogen-based materials of wastes and dead tissues, these reactions are also releasing some of the converted nitrogen compounds directly into the atmosphere as nitrogen (N_2) gas. This is similar to decomposers' release of carbon-dioxide (CO_2) as seen earlier in the carbon cycle. (Notes 26, page 62; Notes 27, page 65; and Notes 28, page 67)

In fact, it should be noted that there are additional types of bacteria that can utilize atmospheric nitrogen gas, turning them directly into nitrates, making for a very efficient system of moving nitrogen through the biosphere and its ecosystems. I need remind my you that all of these nutrient cycles, as well as the flow of energy (E), are going on simultaneously. Remember "…The rivers are our brothers. They quench our thirst…" "…All things are connected…" (Chief Seattle, 1854)

With all this information on plants and how they love a lot of nitrogen to grow, yes, a song has come to mind, look -up and listen to Eric Clapton's, song ♫ *Let it Grow* ♫ (1974). It's really a "hip" tune!

The Nitrogen Cycle: The Human Impacts

The Nitrogen Cycle, like all the other nutrient cycles, can have many human impacts on the surrounding ecosystems that can be positive as well as negative. (Notes 39, page 89) When the cycle is understood and used properly, its processes have become a valuable resource in the agricultural industry. Just ask a farmer why they put animals' fecal and urinary wastes on their farm fields. As previously discussed, the decomposers will break down the nutrient-rich wastes, returning their nitrogen-based elements and compounds to the soil, thereby enriching it for future crop growth. I would ask my students, "has anyone heard of farmers rotating their crops?" Farmers know corn, with its high demand for nitrogen, will drain the soil of large amounts of its nitrogen-based compounds. So, the next year the farmer will plant some type of "legume" (e.g., peas, beans, clover, peanuts, alfalfas) in the field. This is crop rotation – plant one type of crop one year followed by a different crop the next. This is done because legumes love their nitrogen-based nutrients so much they get involved in very special, intimate relationships with bacteria. It's known as a "symbiotic relationship." (Notes 40, page 90)

Yep! Another song that highlights this concept came to mind, right out of the clear blue, you should look-up and listen to a little, ♫ *Muskrat Love* ♫ (1976).

In this relationship both organisms love each other because they both benefit from each other. And legumes cherish this relationship like a muskrat cherishes its musk. Whaaat?

39

POSITIVE IMPACTS FROM N_2 :)

FARMERS — PLACE THEIR ANIMALS'
URINARY AND FECAL WASTES
DIRECTLY ON THEIR FIELDS

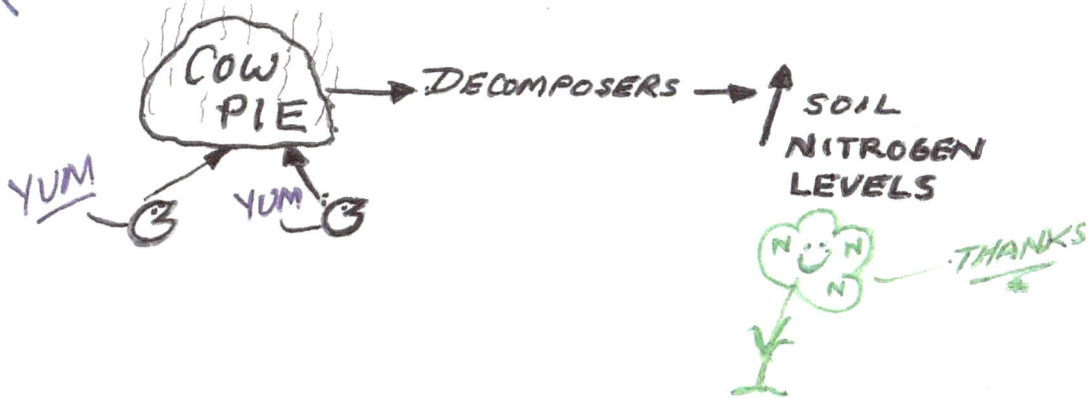

COW PIE → DECOMPOSERS → ↑ SOIL NITROGEN LEVELS

YUM YUM

THANKS

FARMERS — USE NITROGEN FIXING
PLANTS "LEGUMES"

W/ N_2 FIXING BACTERIAS → CREATE
(NH_4)
AMMONIUM

REMEMBER — LIGHTNING
CAN FIX IT "e"

— BACTERIAS
FIX IT YUM

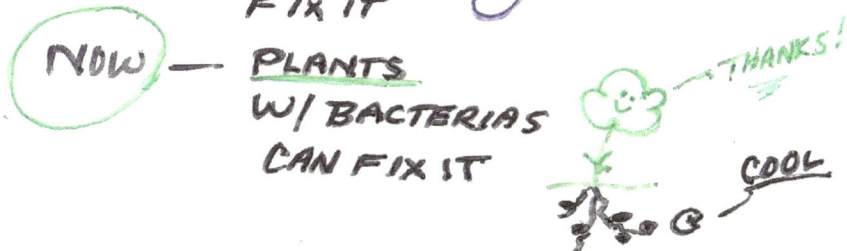

NOW — PLANTS
W/ BACTERIAS
CAN FIX IT

THANKS!

COOL

40 LEGUME — PRODUCER, SPECIAL PLANTZ

Ex. PEAS, BEANS, PEANUTS, CLOVERS

THAT CARRY ON A SPECIAL
RELATIONSHIP w/ BACTERIA

"SYMBIOTIC RELATIONSHIP"

LIVE TOGETHER

— BOTH ORGANISMS
BENEFIT FROM EACH OTHER

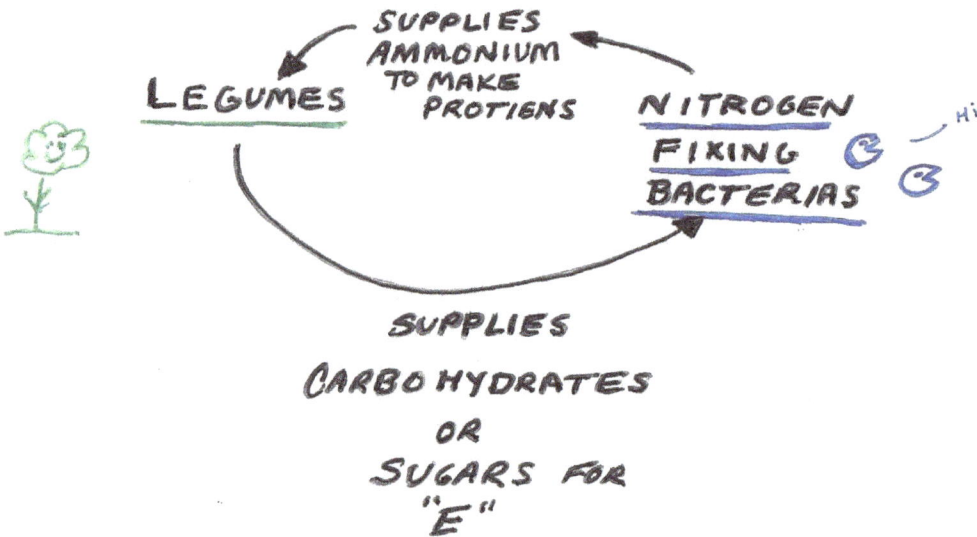

SUPPLIES
AMMONIUM
TO MAKE
PROTIENS

LEGUMES   NITROGEN
                 FIXING
                 BACTERIAS   HI

SUPPLIES
CARBOHYDRATES
OR
SUGARS FOR
"E"

Legumes have this ability to produce tons of sugars, or carbohydrates, over and above what they need to grow and reproduce. However, they can only do this if they receive the required amounts of nitrogen-based nutrients. This is where the nitrogen fixing bacteria become legumes best friend. As outlined on the nitrogen cycle flow chart, the legumes receive the ammonium (NH_4), nitrite (NO_2), and nitrate (NO_3) nutrients from the bacterial action in the soil in close proximity with its root system. These nitrogen rich nutrients aid in the legume's continued growth and reproduction. Almost simultaneously, the legumes supply the bacteria with the sugars and/or carbohydrates they make and utilize during their photosynthesis process. In return, the bacteria receive the energy molecules and compounds needed for their growth and reproduction. As shown in Notes 41, page 92, if one were to dig up some well-established bean plants in the garden, or clover in the yard, one would see the nodules or out-growths of the roots where the bacteria are doing that thing they do - producing lots of nitrogen-based compounds. (Campbell, 1996) In fact, these guys fix so much nitrogen that a good amount is retained in the soil, building up the concentration of nitrates for future soil enrichment. (Notes 41, page 92) The nitrogen cycle, if not understood and left unchecked, can also have negative impacts on our biosphere. These are mainly caused by the release of untreated sewage and/or rain induced runoffs which make their way into the Earth's waterways. As discussed earlier, sewage, or any other organic tissues or particles, are food for decomposers. When potential nutrients from sewage are not removed before their release back into a waterway, the waterway now becomes the environment for biological actions of decomposers, mainly bacteria and single-celled, heterotrophic organisms called protist. The resulting deluge of sewage has many impacts on our waterways as it relates to the nitrogen cycle, or all the nutrient cycles for that matter. When untreated sewage is released, decomposition rates increase rapidly. This means decomposers will feed, grow, and reproduce at alarming rates. For example, given the right conditions, some bacteria can divide their cells, doubling their populations every twenty minutes! (Campbell, 1996)

41

✻ THE PROCESS OCCURS ON
THE ROOTS OF LEGUMES

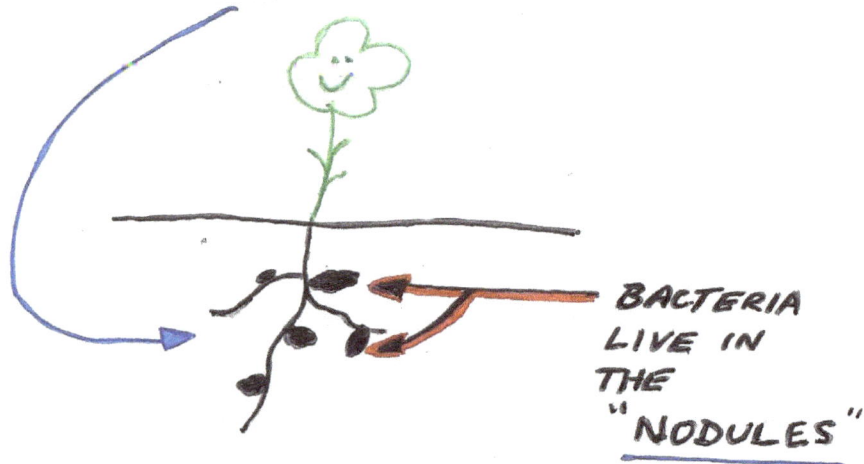

BACTERIA
LIVE IN
THE
"NODULES"

— USED BY FARMERS IN
CROP ROTATIONS

— THE SYMBIOTIC RELATIONSHIP
RESULTS IN EXCESS NH_4
PRODUCTION AND IT'S RELEASED
INTO THE SOIL

Ex.

1ST SEASON	2ND SEASON
CORN	LEGUMES
↓ REDUCES	↑ INCREASES
"N"	N

At this point, I now have to come back to the idea that love is like oxygen - not enough and things will start to die. For example, when sewage or any type of food is dumped, decomposers' populations increase, creating a high demand on the waterway for its dissolved oxygen (D.O.). These microbes grow and reproduce, having even more offspring that strips the waterway of even more dissolved oxygen and the problem compounds itself. These groups of organisms are considered aerobic because they live in the presence of oxygen. Just like most heterotroph (birds, worms, fish, and humans) they use oxygen as a component of their reparation. Yes, dissolved oxygen is a very important gas required by most aquatic organisms. And yes, that includes aquatic plants, algae, and other aquatic autotrophs. As mentioned earlier in the carbon cycle (Notes 27, page 65), the oceans and their surpluses of oxygen that are released make their way from the water and into the biosphere. Fresh water systems are essentially the same, given the right temperature and time of year. Thankfully, like the oceans, they produce more oxygen during the day than they could possibly use up at night, leaving a surplus for other aerobic organisms to utilize.

Oxygen dissolves in water to a limited extent, varying with the pH and water temperatures. These levels of oxygen are critical for much of the aquatic ecosystem's survival. The range for most aquatic, oxygen loving organisms is approximately 3 to 10 plus parts per million (ppm). To give an idea on the implications of this range, I'll use freshwater fish as an example. Three (3) to five (5) ppm supports some fish, but it is a level that is still stressful on most game fish. Rough fish, such as carp and suckers, can tolerate these lower D. O. levels but are still stressed. I would remind my students that this is why a gold fish could survive for so long in that mucky fish bowl on their dressers. I'm sure most everyone knew what I was talking about. Six (6) to seven (7) plus ppm D.O. supports a fair amount of fish, such as pan fish, perch, bass, and northern pike. At 10 or more ppm D.O. levels, an aquatic system can well support an abundant variety of fish, including trout. (Oran, n.d.)

Consequently, the more D.O., the more aquatic organisms the waterway will support. The dumping or release of untreated sewage into a waterway results in a major growth of decomposers as well as other microscopic organisms, such as zooplankton. This rapid growth results in a high demand for D.O. In fact, if the decomposition rates drop the D.O. levels around 1 to 2 ppm fish kills have already started. Plus, other aquatic critters, such as crayfish, snails, insects and their larvae, along with many other organisms, will start to die. Of course, that just adds food to the dinner plates of decomposers furthering the problem of declining D.O. concentrations. Consequently, these high decomposition rates have the opposite effects on the nitrogen levels - they go up. The reason for these increased levels of nitrogen-based nutrients results from the decomposition process discussed earlier. (Notes 38, page 85) The decomposers feed on the dead bodies and wastes of other organisms releasing their nitrogen compounds. These nitrogen compounds become the nutrients that stimulate the rapid growth of plants, algae, and other aquatic autotrophs. This is where some people, perhaps even my readers, start to give that "who's on first…" look.

Most people would reason with increased autotrophic growth, wouldn't that be good for the waterway, with the increased levels of oxygen released? I'd disappoint them by telling them these are unnatural rapid growths. High decomposition rates re-release nitrogen-based compounds that allow algae to grow and reproduction in a very short time. This results in a problem known as "algae blooms." These are extremely high algae and other aquatic plant growths that create disruptions in the natural balance of aquatic ecosystems. In many cases algae blooms are so thick they can make a waterway look like pea soup. Not only do the algae grow rapidly, but so do all the other aquatic plants. As a result, because of the plant and algae growth, the availability for light becomes an issue. What was growing, eventually dies, and decays adding to even higher decomposition rates and loss of D.O. Plus, with the release of nutrients back into the waterway, it promotes additional algae blooms that grow and die, just like a "domino effect," and with each cycle, the problem intensifies.

42

DIVERSITY OF BACTERIAS

✱ THESE KINGDOMS LIVE IN A WIDE RANGE OF ENVIRONMENTS
- FOOD, WATER, VARYING TEMP., AND RELATIONSHIPS W/ ORGANISMS

TYPES OF RESPIRATIONS

A. AEROBIC RESPIRATION — O_2 IS REQUIRED TO BREAK DOWN GLUCOSE TO "E"

I DON'T SMELL A THING

OBLIGATE AEROBES
REQUIRE AIR/O_2
- THESE BACTERIA NEED O_2 TO SURVIVE
- FOUND IN AERATED WATERWAYS AND SOILS

B. ANAEROBIC RESPIRATION — "E" IS PRODUCED W/O THE PRESENCE OF O_2

YIKES *THAT STINKS*

OBLIGATE ANAEROBES
- BACTERIA THAT WILL NOT GROW IN THE PRESENCE OF O_2 (IF EXPOSED FOR A PERIOD OF TIME DIE)

C. FACULTATIVE
- BACTERIA THAT CAN LIVE IN ENVIRONMENTS OF ANAEROBIC AND AEROBIC CONDITIONS

As mentioned earlier, these growth, death, and decay cycles result in much, if not all, of the D.O. in a waterway to become depleted. Yikes! Now the problem compounds itself and becomes even worse when other varieties of decomposers, that do not require oxygen, start to populate. Just like all the parts of the biosphere, ecosystems have a variety of different species of decomposers that, not only can survive in extreme temperature conditions, for instance, volcanic steam ponds found in Yellowstone National Park, but can survive in a range of oxygen concentrations. The aerobic decomposers live in the presence of oxygen; they require it. (Notes 42, page 95) There are others known as facultative organisms, mainly bacteria, that can live in the presence or absence of oxygen, they don't require it, but they can live in the presence of it. However, when the oxygen levels drop off to near zero concentrations, another decomposing bacterium takes over. These guys are called "anaerobic bacteria." These organisms can only live in the absence of oxygen, and these guys are foul to say the least. They take over, breaking down all the dead and decaying plants and algae tissues without the need for oxygen. Wow! Do these guys produce gas? Yes, much worse than a wayside outhouse! With that comes to mind a song called

♫ *Scatman (Ski-ba-bop-ba-dop-bop)* ♫ (1995). This is really a "funky" song – check it out!

These anaerobic bacteria produce a lot of Hydrogen Sulfide (H_2S) that smells like rotten eggs, along with the release of other gases such as Methane (CH_4). When these guys take over a waterway, it becomes even more toxic, furthering the stench from the increased rates of anaerobic decomposition. This has led to the closing of beaches, water stagnation, as well as foul, undesirable odors and tastes to drinking water. Not to mention the loss of the waterway's ecosystem structure. (Notes 43, page 97)

The dumping of untreated sewage is a good example of man's direct impacts that have created many dysfunctional aquatic ecosystems in our waterways where a lot of the aquatic life has died back or been forced to move away because of nutrient cycles gone wrong.

43

NEGATIVE IMPACTS FROM "N"

REMEMBER: "N_2" IS AN IMPORTANT PLANT NUTRIENT

EX.) FOR GROWTH

↑ NO_3 IN AQUATIC ECOSYSTEMS

RESULTS IN
↓

EXTREMELY HIGH PLANT GROWTH "ALGAE BLOOMS"

↓
RESULTS IN
MASS ALGAE DEATH AND DECAY

↓

RESULTS IN

— DECREASE IN D.O., WHICH LOWERS THE ABILITY OF GAME FISH TO LIVE

— WATER STAGNATION, THE SLOWING OF WATER CIRCULATION

— UNDESIRABLE TASTE, ODOR AND APPEARANCE

EX.) FOULS BEACHES AND DRINKING WATER SUPPLIES

This is where I ask the big question, "Where does it go when you go?" (Notes 44, page 99) I would explain this is why developed countries, with their big cities and huge populations turned to the development of wastewater treatment systems. (Notes 44, page 99) In these systems, the sewage is physically and biologically treated using the natural processes of nature. Basically sewage, by the millions of gallons from homes, businesses, and industry, is pumped and channeled through a series of large pipes that flow directly to a wastewater treatment plant. In the first stages of treatment, the sewage is screened to remove larger objects, such as plastics, wood, rocks, etc., that make their way into the sewer system. The water is then slowed down and pumped into the first of two settling tanks where larger particles and debris sink and settle are collected off the bottom, while oils and other debris that float can be skimmed off the surface; both of which are eventually sent to a landfill. From there, the remaining sewage with dissolved particles is sent to a series of aeration tanks. Here, aerobic bacteria and protist can feed on the remaining dissolved organics. Think about it - lots of food, and plenty of pumped in oxygen, what's there not to like for the critters; it's a major Thanksgiving Day feast for them. As the decomposers make their way through the aeration tanks, they act like people after a Thanksgiving Day meal. They settle into a couch to watch a football game; or no, I mean the decomposers settle out in a second series of settling tanks. This is where the now overfed and lazy decomposers settle themselves to the bottom of the tanks, while any remaining organics that float are once again skimmed off. The organisms that settled to the bottom of these tanks are slowly swept off, pumped to more tanks for further biological digesting, and processed into organic fertilizers. The resulting clarified water is sent off for chlorination and discharged to a flowing waterway. (Arms, 2008) It's important to note if we hadn't designed and built wastewater treatment systems to replace open sewers in our city streets in the late 1800's and early 1900's, we probably wouldn't have lasted as a long as we have as a civilization. (Benarde, 1973) If not for these treatment systems, we would be dumping billions and billions of raw sewage daily in to our waterways making them uninhabitable. I would remind my readers to think about what our environment would look like without decomposers. In fact, I would tell them, I think Oliva Newton John recorded a song about it, or something about love, ♫ *If Not for you* ♫, (1971). Listen to it, I think it has some "sweetness man."

44 WHERE DOES IT GO WHEN YOU GO?

TO A WASTE WATER
TREATMENT PLANT

"A BASIC FLOW CHART"
OF A SYSTEM

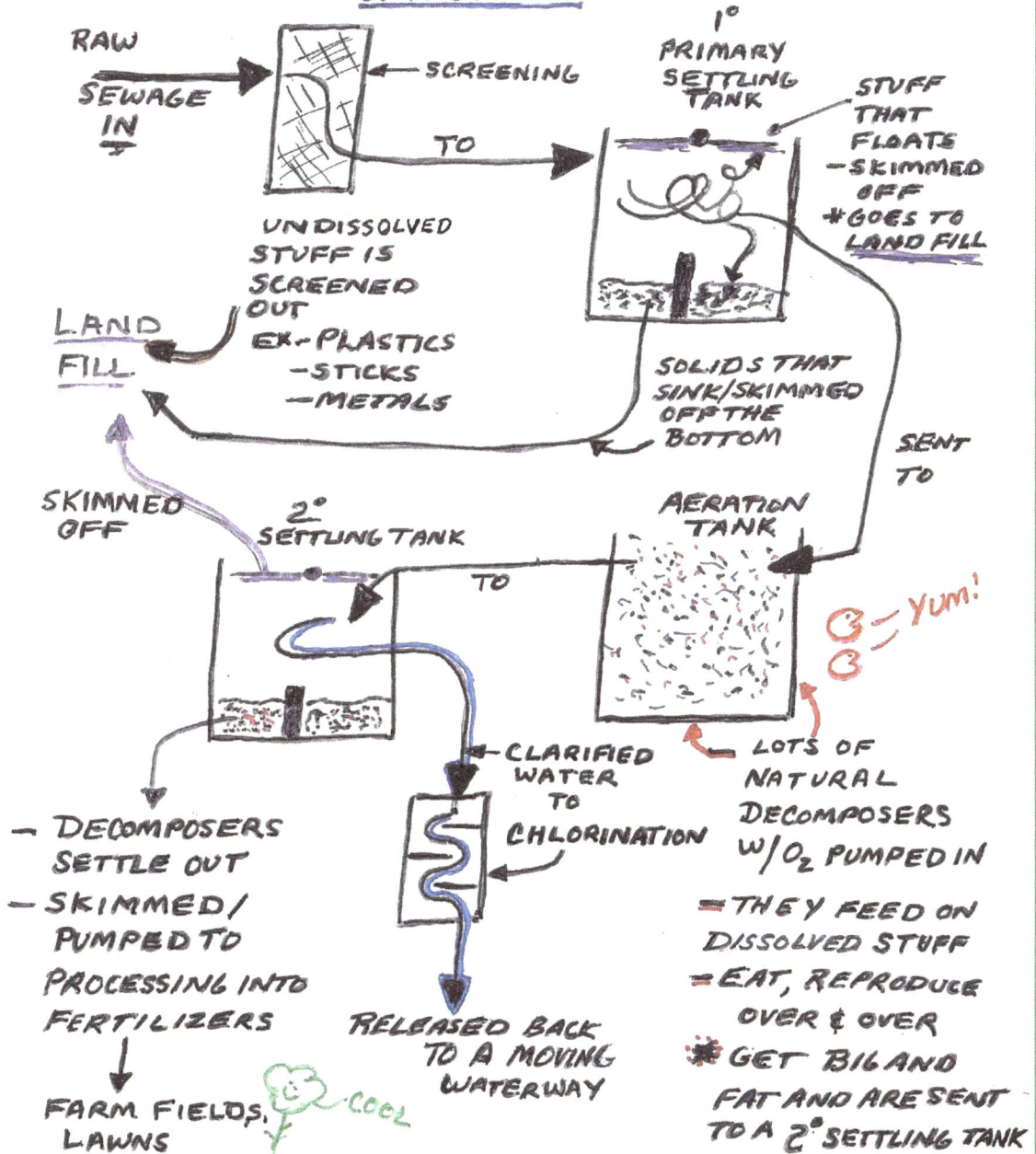

WOW I DID NOT KNOW!

RAW SEWAGE IN → SCREENING

1° PRIMARY SETTLING TANK

STUFF THAT FLOATS
- SKIMMED OFF
※ GOES TO LAND FILL

UNDISSOLVED STUFF IS SCREENED OUT
EX - PLASTICS
- STICKS
- METALS

TO →

LAND FILL

SOLIDS THAT SINK/SKIMMED OFF THE BOTTOM

SENT TO

SKIMMED OFF

2° SETTLING TANK

AERATION TANK

☺ — YUM!

TO →

CLARIFIED WATER TO CHLORINATION

LOTS OF NATURAL DECOMPOSERS W/ O_2 PUMPED IN

- DECOMPOSERS SETTLE OUT
- SKIMMED/PUMPED TO PROCESSING INTO FERTILIZERS
→ FARM FIELDS, LAWNS

COOL

RELEASED BACK TO A MOVING WATERWAY

= THEY FEED ON DISSOLVED STUFF
= EAT, REPRODUCE OVER & OVER
✳ GET BIG AND FAT AND ARE SENT TO A 2° SETTLING TANK

99

SOURCES OF N_2 POLLUTION

DISCHARGES OF RAW SEWAGE

EX. — SEWAGE DUMPED - OR BYPASSES THE
TREATMENT SYSTEM

OR — OVERWHELMED BY HEAVY
RAINS OR MELTING SNOW

— TREATMENT SYSTEM HAS AN UPSET
OR MISFUNCTIONS

WHAT

RUNOFFS — EXCESS WATER FROM HIGH VOLUMES
OF RAIN OR MELTING THAT RUNS
DIRECTLY TO AN OPEN WATERWAY

EX. AGRICULTURAL PRACTICES
— SOILS
— ANIMAL WASTES
— CHEMICAL FERTILIZERS

CITY STREETS
— SOIL/LAWN FERTILIZERS
— DEAD PLANT TISSUES
— TRASH AND GARBAGE
— LEAKING VEHICLES
— MOTOR, HYDRAULIC OILS
— COOLANTS
— PARTICULATES FROM
BRAKES AND TIRE WEAR
— CONSTRUCTION SITE

ALL PROMOTE
ALGAE BLOOMS
AND POLLUTED
WATERWAYS

TO PREVENT RUNOFFS

—SOIL CATCHES AROUND CONSTRUCTION SITES (SILT FENCES)

AGRICULTURAL

- PLOWING PRACTICES
  - AWAY FROM RUNNING WATERWAYS
  - FARM ON LOW SLOPING LANDS
  - PLANT GRASSES ON OPEN SOIL

CITY STREETS

—EDUCATE THE PUBLIC ON WHERE RAIN WATER FLOWS

- STREET SWEEPERS

- WASH CARS ON THE LAWN

- BUILDING OF RETENTION PONDS

HELP!

In addition to untreated sewage as a source of nitrogen-based pollution, there's another source called "Runoff." These are nutrients that are not only causing water quality problems but are very difficult to treat. Runoff comes from many sources. (Notes 45, page 100) It occurs with heavy rains or large volumes of melting snow that cannot be absorbed by impermeable surfaces, such as city streets and sidewalks or saturated and/or frozen grounds. As a result, there's nowhere for these waters to go but run directly into streams, rivers, and lakes. Along the way, these large volumes of water pick up and carry nutrient rich soils from open fields, pastures, and construction sites. The problem compounds itself when these waters pick up animal wastes and/or chemical-based fertilizers that contain high levels of nitrogen as well as other minerals that add to the algae bloom cycle. (Notes 45, page 100)

City streets are also a major concern with runoff. When heavy rains or snow melts occur, they literally wash the streets of many potential nutrients for decomposers and algae to feed on, by picking up dead plant and animal tissues, as well as lawn fertilizers, all of which are carried directly into the surrounding waterway without treatment. If one lives or drives in a farming community, one would notice the way farmers plant their crops on sloping terrain. The crops are planted in strips, going against the grain of the hill, in between each of the strips of crops, grasses are planted to catch the soil particles, preventing their runoff and eventual loss. (Notes 46, page 101) Also, farming practices leave a green area of grass, shrubbery, and/or trees at the edges of crops and animal grazing lands to prevent the runoff of soil and fertilizers.

If one lives in or drives into a city/urban area, one would notice the silt fences around construction sites to prevent soil particles from reaching a waterway. Garbage and other debris are another concern with street runoff. Again, the water picks up cigarette buts, wrappers, cans, bottles, and other assorted garbage that gets flushed directly into storm sewers which then flows untreated directly into a waterway. There is also the notorious leaking of hazardous materials from vehicles such as motor and hydraulic oils as well as coolants. Not to mention the particulates released from the wearing of brakes and tires, all of which again are flushed off the streets and deposited directly in a waterway. Just think about all the cars and trucks on our nation's highways. I would also add that I think Jodi Mitchell (1970) and Counting Crows (2002) got it right with their rendition of ♫ *Big Yellow Taxi* ♫. A relevant song – that "rocks" - turn it up.

To alleviate some of these runoff problems we now see retention ponds that collect and settle out rainwater of its debris and potential nutrients before they reach a natural waterway (Notes 46, page 101). In addition, there is a host of small-scale projects that are having positive impacts on urban waterways. These include everything from street sweepers to public awareness and volunteer programs, all armed to help curb the runoff of unnatural cycling of nitrogen-based and other nutrients along with a variety of debris from polluting our waterways.

Remember "…The shining rivers that moves in the streams and rivers is not just water, but the blood our ancestors…" (Chief Seattle, 1854)

THE PHOSPHORUS CYCLE

47

(D.) | THE PHOSPHORUS (P) CYCLE

3 STORES OF "P"

1.) INORGANIC ELEMENT

"P" ⟶ FROM THE WEATHERING OF ROCKS

2.) INORGANIC COMPOUNDS OF "P"

"P" LOVES TO BOND W/O_2 ⟶ PO_4 PHOSPHATE

JUST LIKE "N" & "C"

✻ FOUND IN THE SOIL OR WATER

3.) ORGANIC COMPOUNDS OF "P"

– FOUND IN LIVING CELLS

EX. PROTEINS FOR TISSUES

NOTE: ALL ARE WATER SOLUBLE

∠ LOOK

OH OH!

Before I start the phosphorus (P) cycle, there are two concepts that I presented before that need to be remembered. First, all these nutrient cycles as well as the flow of energy (E), which will be addressed in the next unit, are all going on simultaneously. Second, one needs to remember the "Law of Conservation of Mass." (Notes 22, page 53)

Like the mineral cycle, phosphorus (P) is released into the environment by way of the weathering process. Once released, it develops into three stores that are important to the biosphere. (Notes 47, page 105) The first store is in the mineral rich-rocks that release the inorganic element form of phosphorus. Once released, the phosphorus acts like many of the other elements in the biosphere. It loves, no I mean it has a deep heartfelt affection for other elements and compounds in the biosphere. This helps phosphorus form a more stable compound or I should say, to form a more stable relationship. And everyone wonders ♫ *What's Love Got to Do With It?* ♫ (1984).

Referring to the phosphorus cycle flow chart (Notes 48, page 107), once released into the soil and/or water system, it essentially acts like "C" and "N." It loves to bond with the abundance of oxygen in the environment to form inorganic compounds of phosphorus called phosphate (PO_4). This becomes the second store of phosphorus in the biosphere. Phosphate formation becomes a particularly important nutrient for most plants and other aquatic autotrophs to grow. Once the phosphates are taken up by the autotrophs, they are fixed into their cellular structures and tissues. This results in the third store of phosphorus in our biosphere, the "organic compounds of phosphorus." (Notes 47, page 107) It becomes the organic form when it's absorbed and incorporated into livings cells. As it moves through the ecosystem, these organic compounds of phosphorus are taken up directly or indirectly by heterotrophs feeding on the plants or the other critters that ate the plants. As a result, the phosphorus compounds are passed along to other living things that use the phosphorus for their growth and reproduction. Eventually, these organic phosphorus compounds are returned to the soil or body of water by decomposers feeding on the wastes or dead bodies of other organisms. And the cycle goes on.

And yes, I have a song ♫ *The Beat Goes On* ♫ by Sonny and Cher (1967). It' "groovy man"- now I'm dating myself.

Nonetheless, remember "…The water's murmur is the voice of my father's father…" (Chief Seattle, 1854)

PHOSPHORUS CYCLE FLOW CHART

THIS CYCLE IS MOSTLY ONE WAY

↳ "LIKE THE MINERAL CYCLE"

48

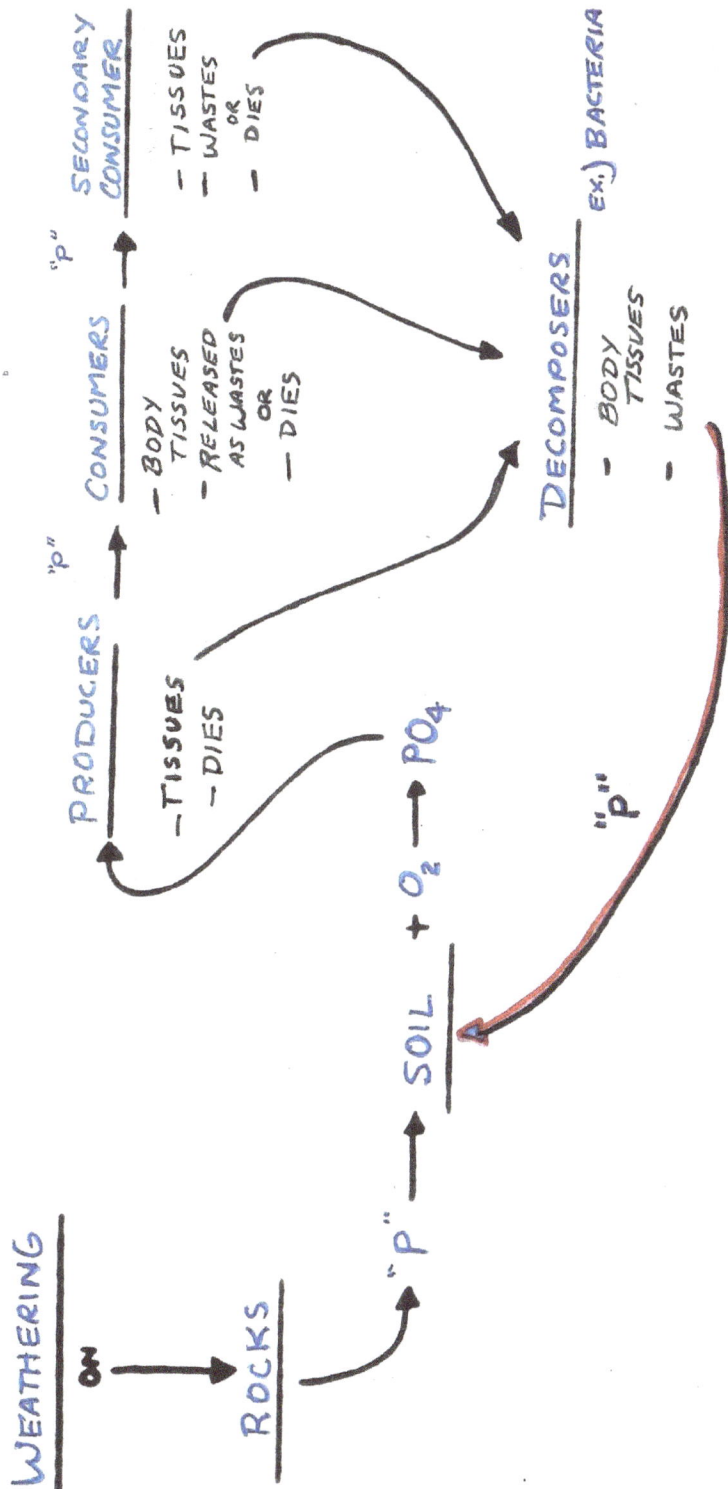

WEATHERING
_____
on → ROCKS
_____

"P" ↗

SOIL + O_2 → PO_4
_____

"P"

PRODUCERS   CONSUMERS   SECONDARY CONSUMER
_____   _____   _____
"P"   "P"   "P"

PRODUCERS
- TISSUES
- DIES

CONSUMERS
- BODY TISSUES
- RELEASED AS WASTES
  OR
- DIES

SECONDARY CONSUMER
- TISSUES
- WASTES
  OR
- DIES

DECOMPOSERS EX.) BACTERIA
_____
- BODY TISSUES
- WASTES

Phosphorus Cycle: The Human Impacts

Many of my students would ask (and probably the reader in this case is also wondering), if this cycle is so similar to the mineral cycle why not lump the phosphorus and its impacts on the environment with those of the mineral nutrient cycle? (Notes 49, page 111) Well, here's why. I would like one to go and find any bottle or container of dish soap or laundry detergent and closely read the label – go ahead. One can finish reading this great section on nutrient pollution after doing that. Go now and do it.

Well, did one find the information I would ask my students to look for? Yes, most of those products are phosphate or phosphorus free except for some dishwasher soaps and garage cleaners. This is because of the phosphate impact on algae blooms. Phosphates are a macronutrient that is required for plants and other autotrophs to grow and reproduce. If one could limit a key nutrient like phosphorus-based compounds from entering our waterways, one could limit or control the growth of algae blooms. This is how phosphorus became known as the "Limiting Factor." (Notes 50, page 112) As an analogy of this concept I would ask my students a question "If we were to be trapped in this classroom with the windows and doors sealed tight, and let's say we had some food and water, how long could we survive?" I'd always get multiple answers. Then my students would quickly reason that the food and water could be rationed, giving us many days to survive. But their love, my love, for oxygen would end up running out. We could have all the food and water we'd ever need, but without the air's 21% oxygen, we wouldn't make it. As a result, oxygen becomes the "Limiting Factor."

Many autotrophs, especially plants, are dependent on a ratio of nutrients they need for proper growth and reproduction. In other words, they require differing types and amounts of micro and macronutrients; of course, this ratio differs on the individual species. It was found that phosphorus stimulates the nitrogen fixing abilities of many algae. A good example is blue green algae, also known as cyanobacteria, a very toxic and poisonous algae that can kill many organisms in an aquatic ecosystem. (Henn, 2016, Sept. 12) With higher phosphorus concentrations, these algae can quickly utilize the available nitrogen-based nutrients to rapidly explode their populations into massive algae blooms turning waterways into looking like "pea soup." I used this example earlier with the nitrogen cycle. (Notes 43, page 97, and Notes 45, page 100) However, just for the sake of simplicity and practicality, let's say in order for algae and other aquatic plants to grow, they require a certain ratio of three macronutrients - carbon, nitrogen, and phosphorus. (Notes 50, page 112) Let's also assume aquatic plants and algae require 50 parts of carbon, to 10 parts of nitrogen, to one-part phosphorus for their normal growth and reproduction. The premise is, if these organisms require this ratio, and if one or more of these nutrients are not available, they will be unable to grow. The carbon and nitrogen portions of the ratio are higher and very difficult to limit or control because their concentrations are all over the biosphere (as shown in the carbon and nitrogen cycles). Consequently, phosphorus is the lowest in the macronutrient ratio and the easiest to control. Why? Because humans are responsible for most of the phosphorus-based compounds being released. We know they are an essential nutrient for plant growth, so we've incorporated them into many of our agricultural and commercial fertilizers. Knowing phosphorus's value as a commercial resource for many products, a whole industry has been formed from the mining, extraction, and processing of phosphorus. This resulted in excess concentrations making their way into many aquatic ecosystems, disrupting the natural nutrient cycles.

As one already knows, because of phosphorus's deep-felt love to interact with other compounds, they make for excellent cleaning agents, most of which ultimately end up being released into our waterways. This is because phosphorus-based compounds are very water soluble. Consequently, even as these phosphorus compounds make their way into a wastewater treatment plant, the biological processes used are mostly heterotrophic decomposers that break up and feed on the dissolved organics. This adds even more to the original concentrations that come into the wastewater treatment plant process. With no autotrophic activities to take up the phosphorus compounds, the decomposition process releases more phosphorus nutrients than they would take up or use. Even after the sewage is biologically treated and clarified, it still contains high levels of dissolved phosphorus-based nutrients that, when released, aid in the growth of aquatic autotrophs. (Notes 51, page 113) This results in large algae blooms that have the same impacts on water quality as seen in the nitrogen cycle with the dumping of raw sewage and runoffs. (Notes 43 - 46) The idea in this situation is to control or stop the use of products that contain phosphorus-based compounds from entering our sewer systems. It will allow us to better control algae blooms and ward off the resulting detrimental effects they have on our waterways. Hence, the ban of phosphorus-based compounds used in most cleaning products, such as hand soaps, dish soaps, laundry detergents, etc.

Additionally, there have been many positive changes in agricultural practices of handling animal wastes and fertilizers, i.e., preventing their runoff. This has helped in the prevention of potential phosphorus-based compounds from running into open waterways, helping to reduce the many water qualities problems we've made. (Notes 50, page 112; Notes 51, page 113)

As always, I would like you to remember the words of wisdom of Chief Seattle, "... We are all connected to the Earth...like the blood that unites one's family..." (1854).

Also, I would like my reader to remember that soaps and detergents may wash out stains on our hands and clothes, but they will never wash away our tears or memories of our ecology. In turn, I would like you to hear the song ♪*A Little Bit of Soap*♪ by the Jarmels (1961). This is a really "hip song." So, make it happen.

THE HUMAN IMPACTS WITH THE PHOSPHORUS CYCLE

REMEMBER: "P" IS A NUTRIENT FOR PLANT GROWTH

LOOK

LIKE "N_2" → CAUSES ALGAE BLOOMS W/THE SAME RESULTS

"PEA SOUP" LOOKING WATERWAYS

* SOURCES OF "P" POLLUTION

* RUNOFF — AGRICULTURAL
   —FERTILIZERS
   — ANIMAL WASTES
   — SOILS

   — CITY STREETS
   - SOIL
   — PLANT TISSUES
   —LAWN FERTILIZERS

* UNTREATED SEWAGE

* INDUSTRIAL WASTES
   — CLEANING AGENTS
   — FOOD PROCESSING

WHAT'S BEING DONE ABOUT "P"

USING "P" AS A "LIMITING FACTOR" FOR PLANT GROWTH

"LIMITING FACTOR"
↳ IS A LEVEL OR CONDITION ABOVE OR BELOW AT WHICH AN ORGANISM CAN LIVE

EX.) – D. O. CONCENTRATION
– TIME
– OR NUTRIENT

LOOK

SIMPLIFIED
↳EX. PLANTS REQUIRE 3 MACRONUTRIENTS

NEEDED IN THIS RATIO

"C" 50 PARTS • "N" 10 PARTS • "P" 1 PART

EASIEST TO CONTROL

* VERY LOW IN THE RATIO

* NOT ——→ AS ABOUNDANT IN THE ECOSYSTEM LIKE "C" & "N" ←— MANY SOURCES

* WE CAN CONTROL ITS LEVELS BECAUSE WE PLACE A LOT OF IT IN OUR WATERWAYS

51

* REMOVE THE "P" AND IT SLOWS THE GROWTH OF AQUATIC PLANTZ AND SLOWS DOWN ALGAE BLOOMS

* STOP THE USE OF "P" AS A CLEANING AGENT

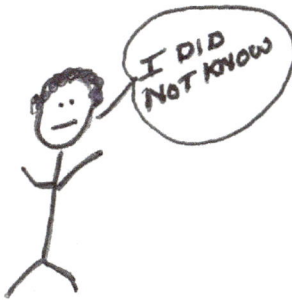

I DID NOT KNOW

EX. DETERGENTS & SOAPS

LOOK ON THE CONTAINER OF THESE PRODUCTS FOR "PHOSPHATE" FREE

* INSTALLATION OF PHOSPHORUS REMOVAL SYSTEMS IN THE WASTEWATER TREATMENT PLANTS

COOL!

REMEMBER "WE ARE ALL CONNECTED TO THE EARTH..."

"LIKE THE BLOOD THAT UNITS ONES FAMILY..."

SECTION THREE

Energy Flow in the Biosphere, a One-Way Connection

(Can you find the "Stick of Dynamite?")

52

ENERGY ("E") FLOW IN THE ECOSYSTEMS

* NOTE

3 REQUIREMENTS
1.) NUTRIENT CYCL.
2.) "E" FLOW
3.) ECOSYSTEM STRUCTURE
   OR PRED.-PREY
RELATIONSHIPS

ENERGY — THE ABILITY TO DO WORK

EX. - MOVE
- GROW
- REPRODUCE
- BUILD — REPLACE OR REPAIR
   CELLS AND/OR TISSUES

YIKES

LOOK

3 FORMS WE NEED TO BE AWARE OF

As the title suggests, the focus of this section will be on the "Flow of Energy in the Biosphere." Before I begin, I would like to do a little review. I did this with my students as a reminder of where we've been, where we are right now, and where we were going. I felt it worked well in the classroom, so I'll start with that thought now. To do this, I will need to return to the example of the Oak tree acorn story. The Oak tree takes up the nutrients from the soil and with the help of the sun, grows and produces acorns, and when planted, will have the right amount of nutrients and energy stored in them to begin the life of another new tree. Keep in mind, that's why the squirrel eats the acorns and the hawk comes down and eats the squirrel. They are all part of the ecosystem structure, all living things require their nutrients, as well as energy to grow and reproduce. Consequently, the ecosystem design lives up to, if one will excuse the pun, its function of allowing life to exist.

Remember, all ecosystems require three common requirements that promote and allow the survival of organism on planet Earth. (Notes 52, page 117) The first requirement is that all ecosystems require the cycling of nutrients. Second, all require the flow of energy (E), which will be the focus this section. And third is the ecosystem structure. We need to remember these are the interactions between the abiotic and biotic portions of the biosphere that enables nutrients and energy to flow through the ecosystem. (Notes 52, page 117) The ecosystem structure is also known as predator-prey relationships, a topic to be covered in more detail in the following sections. As previously done, reminders of concepts already covered will be provided so one can refer to prior notes for clarification. Here's a reminder now. All of these requirements and cycles of eating, growing, reproducing, dying, and decaying are all going on at the same time and for all the same reasons, to allow life to exist. I'm hoping one has gained, and will continue to gain, a greater appreciation of what Chief Seattle meant when he stated, "We are all connected… All things are connected…" (Notes 11, page 30)

1.) SOLAR "E"

ELECTROMAGNETIC
WAVES OF "E"
↓
PROVIDES A SPECTRUM
OF FREQUENCIES

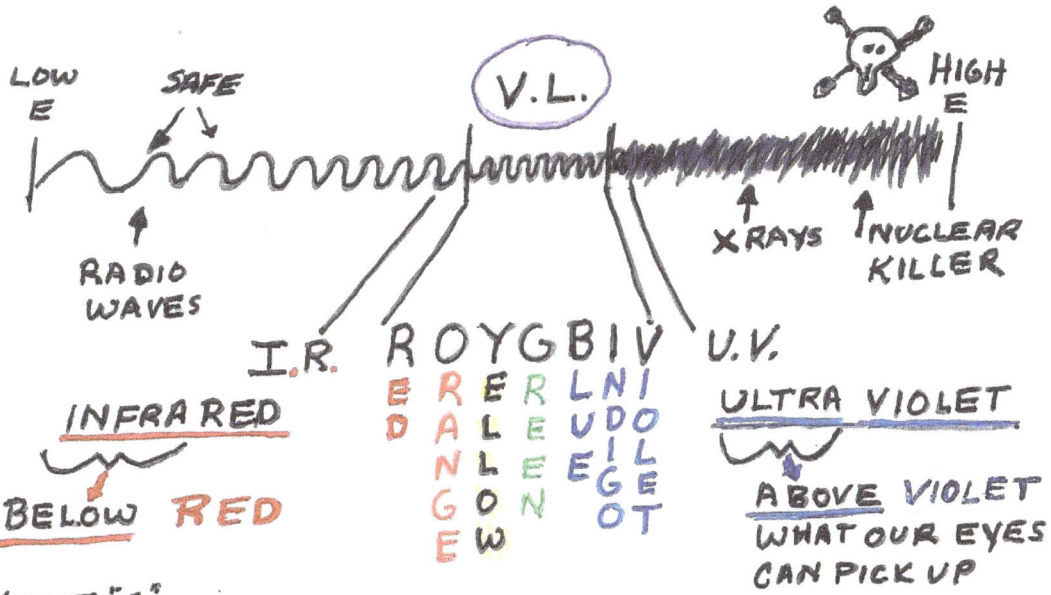

LOW E   SAFE   V.L.   HIGH E

↑ RADIO WAVES   ↑ X RAYS   ↑ NUCLEAR KILLER

I.R.   ROYGBIV   U.V.

INFRA RED   RED ORANGE YELLOW GREEN BLUE INDIGO VIOLET   ULTRA VIOLET

BELOW RED   ABOVE VIOLET
WHAT OUR EYES
CAN PICK UP

- CAUSES
SUN BURNS

HEAT "E"

REMEMBER
THE GREEN
HOUSE
EFFECT

OR USE INFRARED
GOGGLES

I CAN SEE
CLEARLY NOW

119

Energy (E) flow is unique in our biosphere. By definition, energy is the ability to do work. (Petrucci, 1998) (Notes 52, page 117) As it relates to ecology, I like to define it as giving life forms the ability to grow, move, reproduce, etc. For all practical purposes, I like to think of energy as having three forms in our biosphere. The first and most important form is solar energy. Remember what I'd said about the sun earlier? If it were to go out, one would have about eight minutes to get out your ten-pound bag of sugar and set on it, why, "So one can kiss their sweet butt good bye!" And that's truly the case. The sun's energy allows life to perform the necessary functions for life to exist. Yes, energy flow begins with that big red and yellow ball of nuclear power that shines down on our planet. It provides our biosphere with electromagnetic waves of energy. (Notes 53, page 119) It has a wide spectrum with very low energy waves, for example, radio waves on one end that are relatively safe. While on the other end there are very high and deadly energy waves, such as nuclear radiation. Somewhat near the middle is the visible light (V.L.) portion of the spectrum. These are the frequencies we can detect with the human eye, R O Y – G - B I V. (Notes 53, page 119) When looking at my notes, one will see my simple layout of the spectrum. While reading my discussion on its various energy levels, like many of my students, I hope you would find it to be in simple terms, as well as interesting.

If I were asking my students what is infrared, right away many would say night vision goggles, and I would be ok with that thought. I'd also ask them if anyone had seen the movie "Predator" with Arnold Schwarzenegger. (Now I'm really dating myself!) I'd always have a few hands go up. When asked about the part where Schwarzenegger was covered in mud and the predator couldn't see him, what did Schwarzenegger learn? Yes, the predator was using infrared, or heatwaves, for its vision, through its cryptic camouflages. With the mud covering Schwarzenegger's body it blocked his body heat (or infrared energy) from being detected.

"Infra," which means below, are the frequencies/wave lengths just below the range of a human's ability to see red. (Notes 53, page 119) A good example of infrared energy would be the distortion of the background that occurs when looking at a hot grill or a mirage looking effect on a hot road in the middle of summer.

2.) POTENTIAL "E"

— STORED "E"

EX.) SUN'S "E" STORED IN CHEMICAL BONDS

EX.) SUN'S "E" TRAPPED IN ANCIENT TIMES

COAL, OIL, NAT. GAS

3.) HEAT "E"

— THE RELEASE OF "E" THAT RESULTS FROM MOLECULAR MOVEMENT

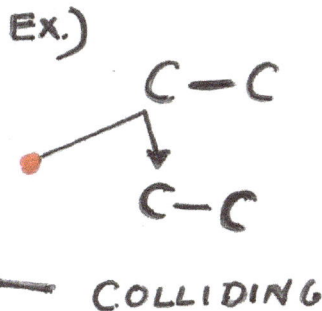

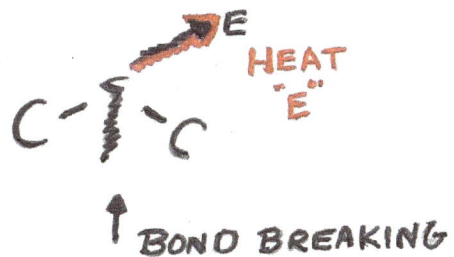

EX.)

C—C
C—C

COLLIDING

C—C HEAT "E"

BOND BREAKING

GREENHOUSE WARMS UP WITH HEAT "E"

On the other end of the visible light range is ultra-violet. "Ultra" means above the frequencies/wave lengths of violet humans cannot see, hence the name "ultra-violet" (U.V.) (Petrucci, 1989). This leads me into discussing sunburns and increased skin cancers that have been attributed to too much skinny-dipping in the cement pond. Many people wonder, how is it that they had been sun burned on a cloudy day? And why does this happen? Well, because, even though most of the U.V. energy is absorbed by our ozone layer, its intensity increases at certain times of the year because of the sun's relationship to the Earths angle. This results in increased levels of U.V. getting through our protective layer. Even with cloudy skies, some breaks through. So, make sure you wear sunscreen with U.V. protection. Just think what it would be like if we didn't have an atmosphere with an ozone layer to protect us? For life on Earth - it would be a "burn baby burn."

As a side note, In the early 1980's scientist found that the Earth's ozone layer, our protective shield, was found to be breaking down. It was discovered that the breakdown was being caused by man's production and use of chlorofluorocarbons (CFC's). Consequently, in 1987, many industrial nations singed an international agreement known as the Montreal Protocol. It called for the reduction and eventual phasing out of CFC's. And, it's been found to be working. The ozone holes discovered on the polar regions of our home are repairing themselves. (E.P.A., 2016) Imagine what could be done with global warming if we had an international agreement to the reduce our CO2 emissions.

Now back to the sun's powerful energy. The sun's energy is the driving force which starts the flow of energy in our biosphere. However, once it strikes the Earth, some of it is reflected back into outer space, some is absorbed by the Earth's ozone as well as its surface. Some is absorbed by producers and/or autotrophs. But what becomes of it once it's absorbed by the photosynthesis process of the producers and/or autotrophs? This is where I would perform my "leaf burning" demonstration. This was a demonstration I used in my classroom as a way to have my students understand where the solar energy is stored and what becomes of it. In this case, I hope one can also benefit from this demonstration by using it as an example. (Notes 55, page 123) To begin I strike a match and light my leaf. Once aflame, I'd waft its aroma towards my nose to take in the beautiful smell, until whoah – I'd drop it as if it had almost burnt my hand (not). I would ask my students', "how did I almost burn my phalanges?" In turn, I would write out our simple photosynthesis reaction on the board by asking my students to tell me what the reactants and products are for the reaction.

BURNING THE LEAF

HEAT

HEAT

MATCH
(ACTIVATION "E")

THERE GOES MY LEAF

BURN THE LEAF

$$CO_2 + H_2O + E \longrightarrow \longrightarrow CO_2 + H_2O + ?$$

$$E?$$

IN

=

OUT

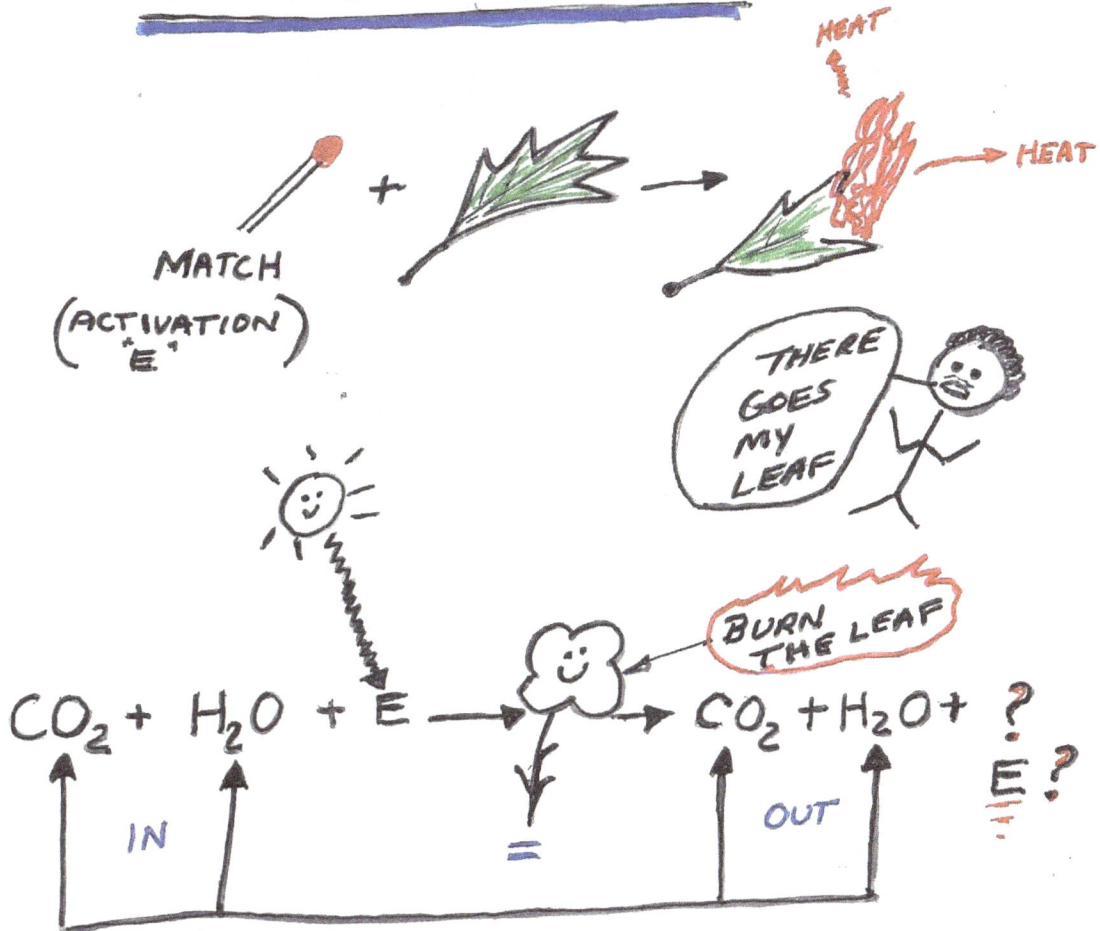

* LAW OF CONSERVATION OF MASS CHECKS OUT

AAAH I'M NOT SURE

* WHERE DID THE FIRE / HEAT COME FROM?

I'd hear "carbon dioxide."

$CO_2 +$

What else? - "Water" good! "Sunshine," or "solar energy," - excellent!

$$CO_2 + H_2O + E \quad \rightarrow \quad \text{🌻} \quad + \quad burn?$$

Producers/Autotrophs

This would complete the photosynthesis reaction the producers and/or autotrophs can utilize to obtain the necessary energy needed for their growth and reproduction. However, I would remind my kids that I just burnt a leaf, a part of a plant, so what just happened to it? This is where my students would give that another Minion look - Whaaat? Again, I would ask them, "I just burnt this leaf, what happened to it?" I would suggest they remember the "Law of Conservation of Mass." What does the law mean? "In" has to equal "out." Then I would quickly get some answers.

$$CO_2 + H_2O + E \quad \rightarrow \quad \text{🌻} \quad + Burn \rightarrow CO_2\uparrow + H_2O\uparrow + Ash.$$

Producers/Autotrophs

I would also add a little of the incombustible material called ash. That, kind of, black stuff that was now on my front counter. I would remind my kids the burning of the leaf is really just part of the carbon cycle. (Notes 27, page 65; Notes 28, page 67; and Notes 29, page 69) I quickly would get the response, "But Mr. Wardinski we're in the Energy section." I would have them look at the reaction again. One should do the same now. We can account for the CO_2 and the H_2O, but where did the energy go? Next, I would say to my students, "hey guys, I almost burned myself on this leaf; how come?" The kids would reply, "because it was hot." True, but where did the energy come from to make it hot? (Notes 55, page 132) No answer. My response was, it's stored in the chemical bonds of carbon that make up the cells and tissues of the leaf. This is known as "Potential Energy," or stored energy. This becomes the second form of energy one needs to become familiar with in the flow of energy in our environment.

Ultimately, when producers and/or autotrophs grow and reproduce, they use the photosynthesis process to absorb, store, and eventual release of the sun's energy. This process continues the flow of energy into the biosphere. Autotrophs, or producers, have this unique ability to carry on the photosynthesis process, to break the bonds of water molecules, swipe the energy from that reaction, and use it to stick carbons together like building blocks, or putting together, "Tinker Toys." Probably few remember, or had the pleasure of playing with Tinker Toys. So, I'd pull out my own set to demonstrate. The wooden discs represented the carbon atoms and the dowels became the chemical bonds that held the disc's together. I'd explain with every dowel I inserted between two wooden discs it was like two carbons being bonded together, not only creating carbon-based tissues, but also storing the energy in the bonds that hold them together. Just think of the "gazillion" carbon bonds that make up all the cells and their inner structures in just one leaf.

Conversely, with the help of a little activation energy from a match, I started the combustion (or burning) process of the leaf. The burning process is the opposite of photosynthesis. In this case, the burning process is breaking the carbon-based chemical bonds that hold together the tissues of the leaf. Or, in the Tinker Toy analogy, the wooden dowels were now being removed. In either case, it results in the release of energy that was stored in the chemical bonds which held the carbon-based tissues together. Once the bonds are broken, they release their energy as heat. This is the heat energy that almost burnt my fingers. Well, not literally, but it was hot! With all this in mind, "Heat Energy" becomes the third form of energy we need to become familiar with in order to understand the flow of energy in the biosphere. This is an addition to solar and stored (or potential) energies covered earlier. (Notes 54, page 121) Heat energy, by definition, is the release of energy that results from the molecular motion of atoms. (Notes 54, page 121) These collisions can be induced by solar radiation striking the atoms of various materials as shown in the Greenhouse effect. (Notes 31, page 75; and Notes 32, page 76) Or, the release of heat energy can result in the making and breaking of chemical bonds. The release of heat energy can be achieved by directly burning the tissues of living things, which was the case with the burning of the leaf, or it can be released by the biochemical reactions found in autotrophs as well as heterotrophs that are either growing, reproducing, eating, or being eaten. As Aldo Leopold once suggested, never assume that one's groceries come from the store and that one's heat comes from the furnace. (Leopold, 1949)

Remember, if we choose to burn the plant material and/or the tissues of the critters that ate the plants, we are heating our homes with the original solar energy that was stored in the chemical bonds of those tissues. In other words, when we burn any organic-based material, we are breaking the chemical bonds that were originally produced by the autotroph's and producer's ability to use the sun's energy in the process of photosynthesis. A great example is the burning of any of the three forms of fossil fuels: coal, oil, and natural gas. (Notes 29, page 69, and Notes 30, page 71) These three fuels are used to run most of the airplanes, trains, and automobiles. Plus, these same energy sources used to stoke the furnaces of industry, to generate electricity and, of course, to heat our homes. This is where my students, and perhaps even my reader, again would give me another Minion "Whaaat" look. And my response is still the same "Yes!" "When you go home tonight, you will be snug and warm knowing your home is being heated from the solar energy of ancient times. It's an interesting, thought provoking, concept, isn't it? Yes, it is, 65 million years in the making! Dead plants and animals that died were buried and left to decompose under great temperatures and pressures, and stored. (Notes 29, page 69 and Notes 30, page 71) As discussed earlier, we are now extracting these fossil fuels for their chemical bonds of energy. Remember the "Carbon Cycle" and its impacts on global warming? (Notes 32, page 76; Notes 33, page 78) As I mentioned before, one can only burn the organic-based materials or tissues directly to release the energy as a form of heat. Once those bonds are broken and energy is released, there's no way to get that energy back.

With this information, I need to explain energy flow in terms of an organism's ability to obtain and transfer energy from the food they consume. My students' usual response to this question was, "from the nutrients we eat." I would answer, "somewhat." Then I would ask them, "do we have these little guys inside us lighting matches or what? What would it be like if we humans, or any other critter for that matter, ate their food only to have it burned up like my leaf?" Let's just say living things, as we know them, wouldn't have made it. I'm sure you're also wondering, how do living things break apart and use the chemical bonds in their food to get their energy? Well "leaf that to me…"

BUILDING MATERIALS FOUND IN NATURE

AIR $\longrightarrow$ CO_2, H_2O, N_2

WATER $\longrightarrow$ CO_2, H_2O, N_2, P, MINERALS

SOIL $\longrightarrow$ CO_2, H_2O, N_2, P, MINERALS

✸ ALL ABIOTIC / NONLIVING PORTIONS OF OUR ECOSYSTEM

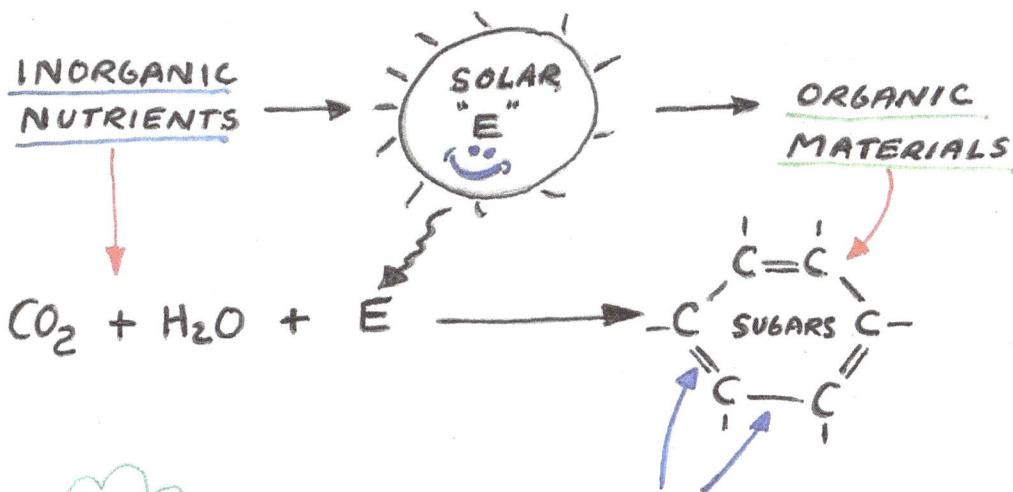

INORGANIC NUTRIENTS $\longrightarrow$ SOLAR "E" $\longrightarrow$ ORGANIC MATERIALS

$CO_2 + H_2O + E \longrightarrow$ SUGARS

THE SUN'S "E" IS TRAPPED IN THE CHEMICAL BONDS OF COMPOUNDS BY WAY OF CHLOROPLASTS

IN TURN, CHEMICAL BONDS FOUND IN FOOD ARE USED BY THE BIOTIC PORTION OF THE ECOSYSTEM

As one may have guessed, there are two sides to every story and the flow of energy in the biosphere is no different. I've simplified this process in my notes. Make sure to reference Notes 57(page 129) and Notes 58(page 130) as I explain the two sides. On one side there is photosynthesis, also known as the producer's side of energy flow. I like to call this the "building up or making of chemical bonds of carbon that can be readily used or stored for later use" side of energy flow. Adding to what I discussed earlier about the photosynthesis process, the producers and/or autotrophs have chloroplasts that contain pigments of chlorophyll in their cells. These are the cellular structures that begin the process of capturing and storing the solar energy so it can be used and passed through the ecosystems for organisms to survive and reproduce. Simply, these guys, with their chloroplasts and specialized pigments, can absorb the sun's energy to synthesize, or make chemical bonds of carbon. These carbon bonds eventually develop into the cells and tissues of producers and other autotrophs. In turn, with further growth and development, these organisms start storing potential energy in their cells, tissues, and especially in the production of carbohydrates, such as sugars, starches, and cellulose. (Notes 57, page 129)

On the other side of energy flow is reparation, also known as the "consumer side" of energy flow. I like to call it the "breaking of chemical bonds for the release of energy" side. What's really cool about both sides of the energy flow is that each have their own unique and specialized catalysts that catch or harness the energy of the sun directly, or from the processes of breaking chemical bonds. With each side having these specialized catalysts it allows them to transfer and/or use the energy without burning it up in a fire ball and lost as heat, which was the case with my leaf.

PRODUCERS
(MAINLY PLANTZ)

CONSUMERS

PHOTOSYNTHESIS SIDE

RESPIRATION SIDE

LIGHT   TO MAKE

- USES CHLOROPHYLL
TO TRAP AND STORE "E"
IN CHEMICAL BONDS

- BREAKING DOWN OF
GLUCOSE TO RELEASE
"E" FROM CHEMICAL
BONDS

- GLUCOSE
(SUGARS)
- TISSUES

← ORGANIC

SOME "E" IS
LOST AS HEAT "E"
TO THE ENVIR.

"E" LOST AS HEAT "E"
TO THE ENVIR.

MAKE
CHEMICAL
BONDS

BREAKING
CHEMICAL
BONDS OF
ORGANIC
COMPOUNDS

$CO_2 + H_2O + E$

(INORGANIC)

$CO_2 + H_2O$

(INORGANIC)

PLANTS USE TO
CARRY OUT LIFE
PROCESSES

EX.) - GROWTH
- REPRODUCTION

ANIMALS EAT
THE PLANTS TO
CARRY OUT LIFE PROCESSES

EX.) - GROWTH
- REPRODUCTION
- MOVEMENT

129

VS.

PRODUCERS SIDE OF "E" FLOW	CONSUMERS SIDE OF "E" FLOW

— USES CHLOROPHYLL
   A PIGMENT USED
   TO TRAP LIGHT
   "E"
	— DO NOT HAVE
	   CHLOROPLASTS

— BONDS (FIXES) INORGANIC NUTRIENTS (H_2O, N_2, P, C, O_2) INTO ORGANIC COMPOUNDS USING THE SUN'S "E"	— BREAK DOWN ORGANIC COMPOUNDS TO RELEASE "E" FROM CHEMICAL BONDS
— "MAKING BONDS" BY CAPTURING "E"	— "BREAKING BONDS" TO RELEASE "E"

✱ AUTOTROPHS
   — MAKE THEIR OWN
FOOD
	✱ HETEROTROPHS
	   — CANNOT MAKE
	   THEIR OWN FOOD

CONSUMERS GET THEIR "E"
DIRECTLY OR INDIRECTLY FROM THE SUN

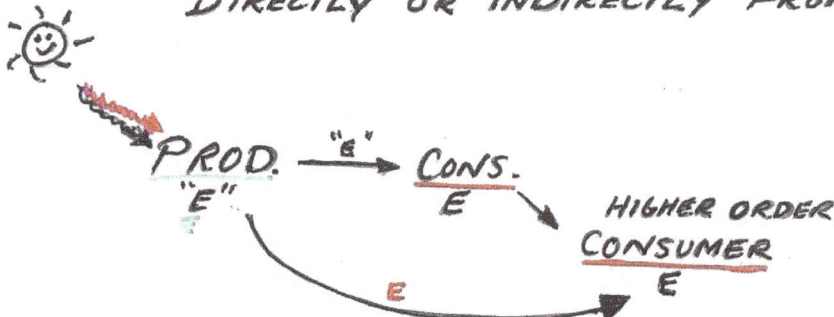

PROD. ——"E"→ CONS. → HIGHER ORDER
"E"   E   CONSUMER
              E   E

I would demonstrate this concept to my students with the "kaboom or firecracker" parable. I would like to use my classroom narrative once again so that you can gain a simpler understanding of how all organisms can transfer and/or use energy for growth and reproduction. I would begin this example by I acting like I'm lighting a firecracker, pretend to throw it, and yell "Kaboom" as loud as I possibly could. That would wake them up! I'd explain the firecracker released all of its energy at once, just like when I burnt the leaf. All the energy was lost as heat. This is where the specialized catalysts for each process come into action. These catalysts have the ability to slow down the photosynthesis and respiration processes, allowing them to control the flow of energy. By controlling these processes, it allows cells to capture and/or release the energy obtained without burning up the chemical bonds into flames only to be lost as heat energy. To further simplify and explain the actions of these catalysts, I would tell my kids to think of the catalysts as Energy Police. (Not the ♫ *The Dream Police* ♫ the song by Cheap Trick, 1979, but the Energy Police.) They have the ability to grab the energy from the firecracker before it's all lost as heat, in one big kaboom. In turn, I would repeat my firecracker kaboom with the Energy Police in action. (This is where I'm nominated for an Academy Oscar award, in the category of "slow-motion theatrics.") Starting out loud and speaking very slowly, I would progress to an ever slower and softer voice, ending with a soft "ooooom," or a "KAaaaabooooom." Then I would ask the class, "What happened to the kaboom or firecracker?" They would reply, "it was grabbed." Yes! The Energy Police grabbed it and is now using it for a chain gang to break rocks for a railroad. HA! I would tell them that was a very good answer. But whaaaat? I'd get the rolling eyes of the class. "Wait," I would tell them, "there is another part to this controlled process used by these catalysts." I would ask, "Did the Energy Police catch all the energy?" Individually, some would say no. And, I would quickly ask, "Are you so sure? How do you know?" "Yes, Mr. W," they would say," because you slowed the kaboom so that most of the firecracker's kaboom was caught by the Energy Police. The students would add they could still hear some soft ooooom at the end. My reply again was, "very, very good, but where did the ooooom go?" I would explain that it was lost to the environment. Just like the firecracker, even though living things have these catalysts that control and utilize the transfer of energies they are not 100% efficient. In other words, during the reactions of photosynthesis and reparation, both eventually give up or lose some of the energy as heat. Or, in my case of the firecracker example, a soft "ooooom" was lost.

Just remember, while the photosynthesis and respiration processes are similar, both are quite different in their energy outcomes. Both are similar by the fact that they both deal with energy and have specialized catalysts for the control of energy transfers, plus the very important fact they both occur at the cellular level. However, that's where the similarities end; otherwise, they are the complete opposite of each other. One carbon at a time, the photosynthesis side of energy flow uses its unique catalysts to make and store the chemical bonds of energy. While on the respiration side of energy flow, other specialized catalysts are utilized to break the chemical carbon bonds for the release of energy. Again, one carbon at a time. (Notes 57, page 129; and Notes 58, page 130)

Hey, I just thought of another song that's fitting for this topic. Have you ever heard the song ♫ *One Piece at a Time* ♫ by Johnny Cash (1976)? A catchy/cool tune - to cool off from all this heat energy.

With that song in mind one carbon at a time and the use of specialized catalyst I continue with my narrative on the reparation process by asking everyone to feel their arms. And I would now have you, my reader, to do the same. I'd ask if it was warm; most would say yes, and I would reply "ok you're alive." Do you know your bodies normal temperature?" If your answer was 98.6 degrees Fahrenheit your right on! Oh-yes there is another groovy tune that coincides with our body heat - I'd start singing the song but it would be better if you just looked it up; ♫ *98.6* ♫ by Keith (1967).

Our cellular respiration process provides us with a stable body temperature. If one gets cold, our brain signals our muscles to start contracting which makes us shiver; or, if one becomes too warm, one starts to sweat. Some organisms don't have that choice; they just shut down, hibernate, or die. In any case, energy is being lost as heat. For example, all organisms carry on the respiration process, ultimately losing some of their energy as heat. And yes, even plants and/or autotrophs utilize the respiration process. Remember when I discussed the oceans, tropical and boreal forests providing us with a reserve of the oxygen, that stuff we so much love? At dusk, when solar light drops, the photosynthesis process stops and the respiration process takes over, keeping their cells energized until dawn. Consumers and/or heterotrophs carry on the respiration process constantly to keep their cells functioning. Some organisms are more efficient at their energy flow than others. I would tell my kids about my compost pile. It's a good example of the heat energy lost by the decomposers chowing down on the organics. There were times when it would become so biologically active from the decomposition process that I couldn't hardly place my hand in it because it was so hot, plus I could smell the release of ammonium compounds of the nitrogen cycle coming off the pile. And just like humans or any other heterotrophic or autotrophic organisms, they all lose varying amounts of energy as heat, to their surroundings.

I would remind everyone about the energy spectrum I had covered earlier. (Notes 53, page 119) If one were to put on a pair of infra-red goggles, one would see a variety of colors and their intensities that result from the variations of heat energy being lost by the tissues of different organisms being examined. If one does not believe me, just call up Arnold Schwarzenegger and ask him. Nonetheless, both sides of energy flow lose some energy as heat; in fact, they both lose a lot. Looking at Notes 57 (page 129) the photosynthesis side of energy flow uses the reactants of CO_2 + H_2O + Solar E to build organic compounds of carbohydrates, such as sugars, starches and cellulose, as a storage of energy. Notice the respiration side of energy flow where the carbohydrates are being broken down to release their stored energy. Also, notice both sides lose a certain amount of energy as heat in their respective processes.

In turn, I would ask my kids, and now I would ask you "does the 'Law of Conservation of Mass' work in this case?" Again, the answer was "yes, both sides have equal amounts of CO_2 + H_2O." "Or wait," I would ask, "are both sides equal or did I forget to write the "E" symbol?" **"No, the flow of energy is only one way in our biosphere – in and only in."** Unlike the cycling of nutrients, energy is the one required component for all living things, that cannot be recycled in the environment. Remember, just like the "ooo$_{ooo}$" left over from the "Kaboom," it's lost.

This leads me to explaining two very, and I mean very, important "Laws of Thermodynamics." These are two laws of energy flow I've been leading up to. They need to be understood before one can appreciate the advantages and disadvantages of using non-renewable and renewable energies, a topic I'll be covering in more detail shortly.

Leak

"E" FLOW IS ONE WAY ⟶ IN ONLY

- LOST TO THE ECOSYSTEM AS HEAT "E"

- CANNOT BE RECYCLED LIKE NUTRIENTS

1ST LAW OF THERMODYNAMICS

"E" IS NEITHER CREATED NOR DESTROYED

THOUGH IT MAY BE TRANSFORMED FROM ONE FORM TO ANOTHER

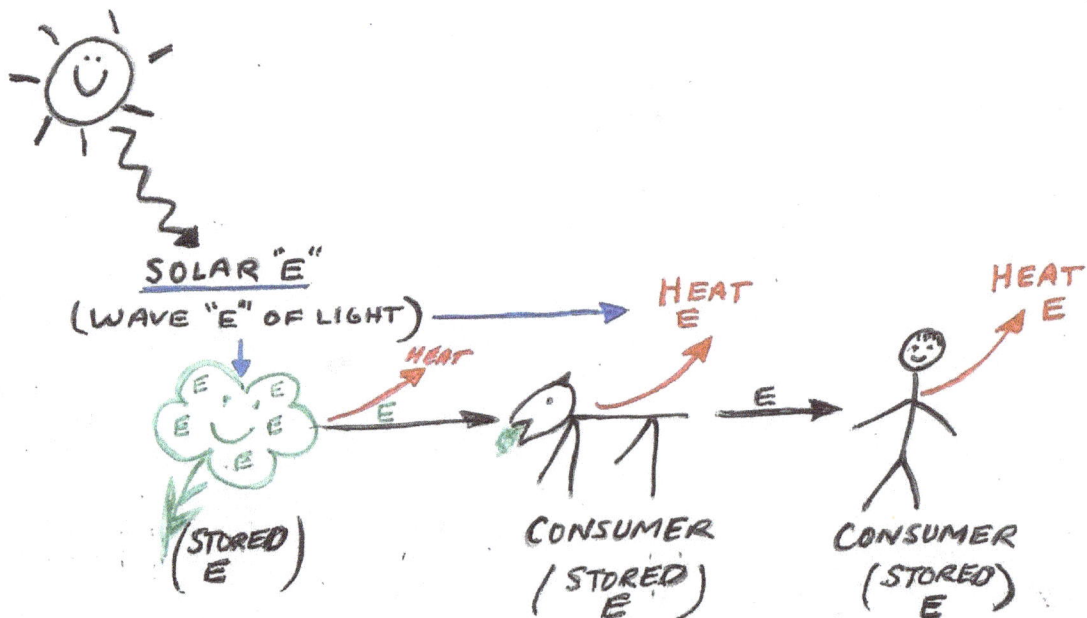

SOLAR "E"
(WAVE "E" OF LIGHT)

HEAT E

HEAT E

(STORED E)

CONSUMER (STORED E)

CONSUMER (STORED E)

The "First Law of Thermodynamics" states energy can neither be created nor destroyed, but it can be transformed from one form to another. (Notes 59, page 134) (Petrucci, 1989) This is a very important law of nature and it has big implications on the survival of living things. What the law is saying is that only the sun can give us energy, and nothing on or in our planet can make energy. Only the autotrophs can capture the sun's energy to make chemical bonds possible. And, even in the process of making them, some of the solar energy is being transformed into heat energy which is lost to the environment. In turn, the heterotrophs just transform the energy into more chemical bonds, at the cost of losing a lot of it as heat. So, either directly or indirectly, all living things receive their energy from the sun.

To review this concept, I'd play a game with my students I called, "Trace the flow of solar E." I'd ask them, "Well go ahead, make my day. Give me an example of something humans have made without the sun's energy?" One response I received was "Mr. W, my desk." I replied, "That's easy. Where did the Iron come from to make the frame? Iron ore that was extracted from the Earth by people or machinery. If people are digging the iron up, I bet they would be very hungry. And, I bet the food they ate came from plants and/or animals that ate the plants, that received their energy from the sun. I'd also bet they were sweating a lot doing that kind of work. It would be no different if one were to use machinery to extract the iron ore. The equipment would be made from and run on fossil fuels, which are nothing more than dead plants and animals that obtained their energy directly or indirectly from the sun."

I would add that I hadn't even touched on the manufacturing of the steel frame or the wood fibers that make up the desk top. Whether the energy is in the form of chemical bonds, heat, electrical, or stored as potential energy, one needs to remember that energy cannot be created, nor destroyed, but only transformed into different types of energy.

In turn, this leads me, very nicely, into the "Second Law of Thermodynamics." (Notes 60, page 137) This law states that each time energy is transformed, it moves from a more concentrated, more organized form, to a less concentrated, less organized form. (Petrucci, 1989) What this law is saying, as energy moves through an ecosystem, less and less energy becomes available to do useful work. In other words, the total amount of energy is still the same (remember the first law that energy cannot be created nor destroyed, only transformed), but the amount converted or transformed to chemical bonds for cellular functions becomes less and less with each conversion. Consequently, while some of the energy is being passed on to other chemical bonds, much of the total energy that originated from the sun is lost as heat to the environment with each transformation. As stated earlier, the photosynthesis and respiration processes are not 100% efficient; in fact, far from it. Each time energy is transformed from one form to another, about 90% is lost as heat. Think about it in terms of individuals eating or being eaten in an ecosystem. With every conversion, 90% of the energy value is lost as heat. Simplified, the law states that less and less energy is available to do useful work, and at the end of the line, all of the original solar energy is lost as heat to the environment. Yes, it is gone!   And one will never ever, ever never, never get it back! It's lost as heat! (Notes 60, page 137)

Ok, I have another timely song, ♫ *End of The Line* ♫ by the Traveling Willburys (1988). These guys could jam – so "I'll catch you on the flip side."

2ND LAW OF THERMODYNAMICS

— FOR EACH TIME "E" IS TRANSFORMED, IT MOVES FROM A MORE CONCENTRATED, ORGANIZED FORM TO A LESS CONCENTRATED, LESS ORGANIZED FORM

OR

AS "E" MOVES THROUGH THE ECOSYSTEM, LESS AND LESS "E" BECOMES AVAILABLE TO DO USEFUL WORK

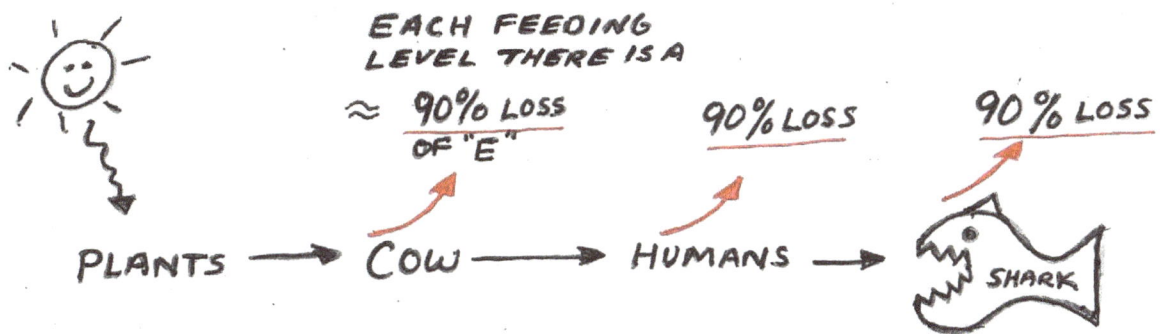

EACH FEEDING LEVEL THERE IS A
$\approx$ 90% LOSS OF "E"   90% LOSS   90% LOSS

PLANTS → COW → HUMANS → SHARK

— THE FOOD CHAIN?
— FOOD REQUIREMENTS?
— FOSSIL FUEL USE?

Energy Flow in the Biosphere: A One-Way Connection, The Human Impacts

Understanding the flow of energy in the biosphere, where it comes from, how producers and consumers are involved, as well as the two laws of thermodynamics, should make it easy to understand and better appreciate the impacts humans have on energy. The main source of energy consumed in most industrialized countries is fossil fuels: coal, oil, and natural gas. (Notes 66, 67, 68) While nuclear power is generally used in the production of electricity, many countries are phasing it out. On the other hand, fossil fuels still play a significant role in the consumption of energy in a variety of ways. They are used not only to generate electrical power but play a leading role in the industrial production of goods, both of which rely heavily on coal and oil. Fossil fuels are still a key factor in the transportation industry. Planes, trains, trucks, and automobiles are dependent on fossil fuels. Oil not only lubricates the engines of most transportation systems but also provides the necessary fuels to run them. While industry has a high demand for electricity, some transportation systems, such as subways, some trains, and cars, also use large amounts of electrical power produced mainly by coal. However, while changes are occurring, much of the world's industries and transportation systems are run directly or indirectly on fossil fuels. (Notes 63, page 142)

While some countries, such as Norway, Sweden, and Germany, are using less fossil fuels every year, the world-wide use is still high. (11 countries, 2019) These countries are increasingly turning to renewable energies to meet their energy needs. (Notes 62, page 141) This is where I would say to my students, "looks like this is a defining moment," as it pertains to renewable and nonrenewable energies. (Notes 62, page 141; and Notes 63, page 142) Renewables are defined as those that can be replaced by natural processes. For example, biomass, water power, wind, and solar, are all essentially inexhaustible. (Arms, 2008) Remember, all energies come by way of the sun, even renewable energies. It's true, the sun's energy can be used directly for heating as seen with the greenhouse effect, or it can be used directly in the generation of electricity with the use of solar panels, a concept I will detail later.

Nonetheless, most of the renewables we use today are driven by the sun's heating and cooling of the Earth during its daily rotation. In fact, if one additionally calculates in the Earth's angle and its elliptical orbit around the sun, there are no two days of a year where the sun's energy strikes the Earth with the same intensity. This variation in heating and cooling sets up weather patterns that deliver the winds, rains, and snows that eventually melt, to create and restore new energies every day. Hence, the name "renewable energies." (Notes 62, page 141)

Nonrenewable are completely the opposite. They are energies that can be used up completely, depleted to the point where they're no longer economical to obtain. (Arms, 2008) In other words, as I tell everyone, "when they're gone, they're gone." (Notes 63, page 142)

There are essentially two nonrenewable energies being used - - fossil fuels and nuclear fuels. As I stated earlier, nuclear fuels are used mainly in the generation of electricity. While fossil fuels are used heavily in the generation of electricity, they also help run much of the industries as we know them. Fossil fuels play a big part in our daily lives providing us with fuels that are used to heat our homes and power our transportation systems, as well as in the production of goods, all of which have made our lives much easier. Remember the sun's energy is stored in the chemical bonds of fossil fuels. Fossil fuels, which are made up of dead plants and animals, have fixed amounts of energy. In other words, there is only so much coal, oil, and natural gas available. (Notes 29, page 69 and Notes 30, page 71) Once these energies are used and their chemical bonds are broken, the energy released is eventually lost to the environment as heat. (Notes 57, page 129 and Notes 60, page 137) Hence, the name, "nonrenewable." Remember, energy flow in the biosphere is one way – in and only in. The transportation and industrial processes I mentioned earlier consume a large portion of the fossil fuels we use today. However, the generation of electrical power has an even higher demand for fossil fuels. That is why I chose the production of electrical power to demonstrate how nonrenewable and renewable energies are being used. I also picked this example because of the ever-increasing demands for electrical power by the world's growing population. For instance, more and more electrical energy is going to be needed to light homes, run industry, power all the assorted electrical devices in use, not to mention the charging of electric cars. Again, the generation of electrical power will make for a good demonstration of energy flow from the human point of view. Plus, I hope it will aid in understanding the human impacts of energy and how they can be minimized by working with the natural laws of energy flow.

NONRENEWABLE VS. RENEWABLE "E"

NOTE: THESE FORMS OF "E" ARE ALL (IN SOME FORM) DIRECTLY RELATED TO SOLAR "E"

RENEWABLE "E"

- REPLACEABLE BY NATURAL PROCESSES
- ESSENTIALLY INEXAUSTABLE

EX.) BIOMASS → WOOD

HYDRO → WATER POWER

ANIMAL → WORK HORSES

} USED BY AGRICULTURAL SOCIETIES

SOLAR

WIND

HYDRO

GEOTHERMAL

} BEING USED MORE BY INDUSTRIAL SOCIETIES

To gain a better understanding of how electrical power is produced, I'll start by using nonrenewable energies. (I'll be referring to Notes 64, page 144, and Notes 65, on page 145, so be sure to look at them as we move along.) For each of the energies being used to generate electrical power, I'll provide some examples of their advantages and disadvantages.

Most of the electrical power generated begins by heating water into high pressure steam. As a review, do you remember where the energy comes from to heat the water into steam? If your response was "from the energy stored in the chemical bonds of dead plants and/or animals that ate the plants and became the fossil fuels we now burn," that would be correct. (Notes 29, page 69 and Notes 30,

NONRENEWABLE

- CAN BE USED UP COMPLETELY
"DEPLETED" - NO LONGER
ECONOMICALLY
FEASIBLE TO OBTAIN

FOSSIL FUELS

COAL

PETROLEUM (OIL)

NATURAL GAS

NUCLEAR FUELS

USED BY INDUSTRIAL SOCIETIES EX.) U.S. CANADA

3 MAIN USES FOR NONRENEWABLE "E"

USES	TYPE OF "E" USED
1.) TRANSPORTATION	- OIL
2.) PRODUCTION OF ELECTRICITY	- MOSTLY COAL - NUCLEAR POWER - SOME NATURAL GAS AND OIL
3.) INDUSTRIAL	- MOSTLY OIL AND COAL

page 71) Yes, just like the burning leaf example and its release of heat energy.

Fossil fuels follow this example very well. By burning fossil fuels, the stored energy is released to heat the water into high pressure steam. The steam is released into the turbine. (Notes 64, page 144 and Notes 65, page 145) The turbines can vary in size, but many turbines used in steam-generated power are massive in size and are comparable to the size of a three-car garage. When the steam is released, it expands out and turns the blades which are mounted on the shaft that runs through the turbine. These blades are comparable to airplane propellers except the turbines contain many more blades with varying sizes to maximize the expansion of the high-pressure steam. From there, the turbine's shaft turns the generator. (Notes 65, page 145) In order for you to gain a general understanding of how electricity is produced, I'll explain the process in its simplest form. The generators used in most power plants are as large as the turbines that turn them. However, their internal components are very different. Inside of the generator are massive windings of wire that when rotated create large magnetic fields that are quickly produced and broken at very rapid speeds. This on and off cycling produces pulses of alternating current, or electricity, onto the grid. These pulses of electrical energy are comparable to a strobe light, which many people are familiar with. In other words, the flashes of light are comparable to the pulses of electricity being released on the grid. Further, if the speed of the strobe light is dramatically increased, it would appear as though the light is continuously on. Pulses of electrical power are similar, they're called "cycles per second." These cycles happen so fast that the power is always flowing. Once on the grid, the electrical power is ready for use by industry, businesses, homes, etc. "Grid" is a term used to represent the system of wires that connect an area's homes, businesses, and industry with the flow of electrical power. It's very similar to the horizontal and vertical lines that make up the squares on a spread sheet or a sheet of graph paper. In either case, once the power is supplied to the grid, lights come on, machines run, subways operate, and electric cars get charged.

Nuclear fuels, another nonrenewable, perform the same process of making pressure steam to generate electrical power. The main difference is that instead of breaking chemical bonds that hold atoms together to release energy, scientists have learned how to break the nuclei of atoms which releases enormous amounts energy. I would like to remind you about the detonation of atomic bombs like those used on

PRODUCTION OF ELECTRICITY

64

NON RENEWABLES
1.) FOSSIL FUELS
2.) NUCLEAR ENERGY

RENEWABLES
1.) BIOMASS

PHOTOSYNTHESIS
$E + H_2O + CO_2$

90%

LOST AS HEAT

90%

90%

DEAD PLANTS
&
ANIMALS

FOSSIL FUELS
- COAL
- OIL
- NATURAL GAS

BOILER HEATS H_2O

STEAM

90%

2.) GEOTHERMAL
LAND   HEAT

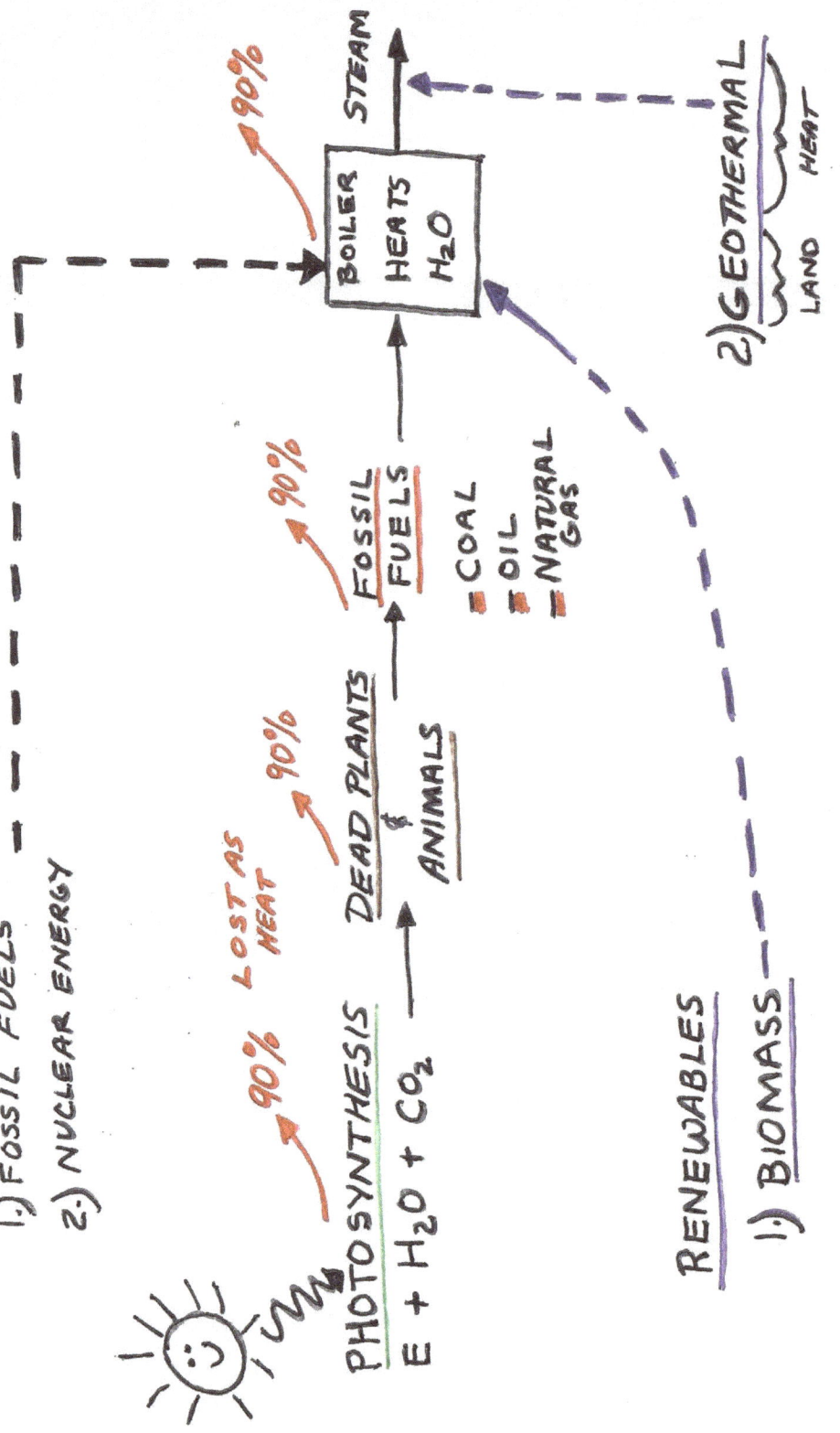

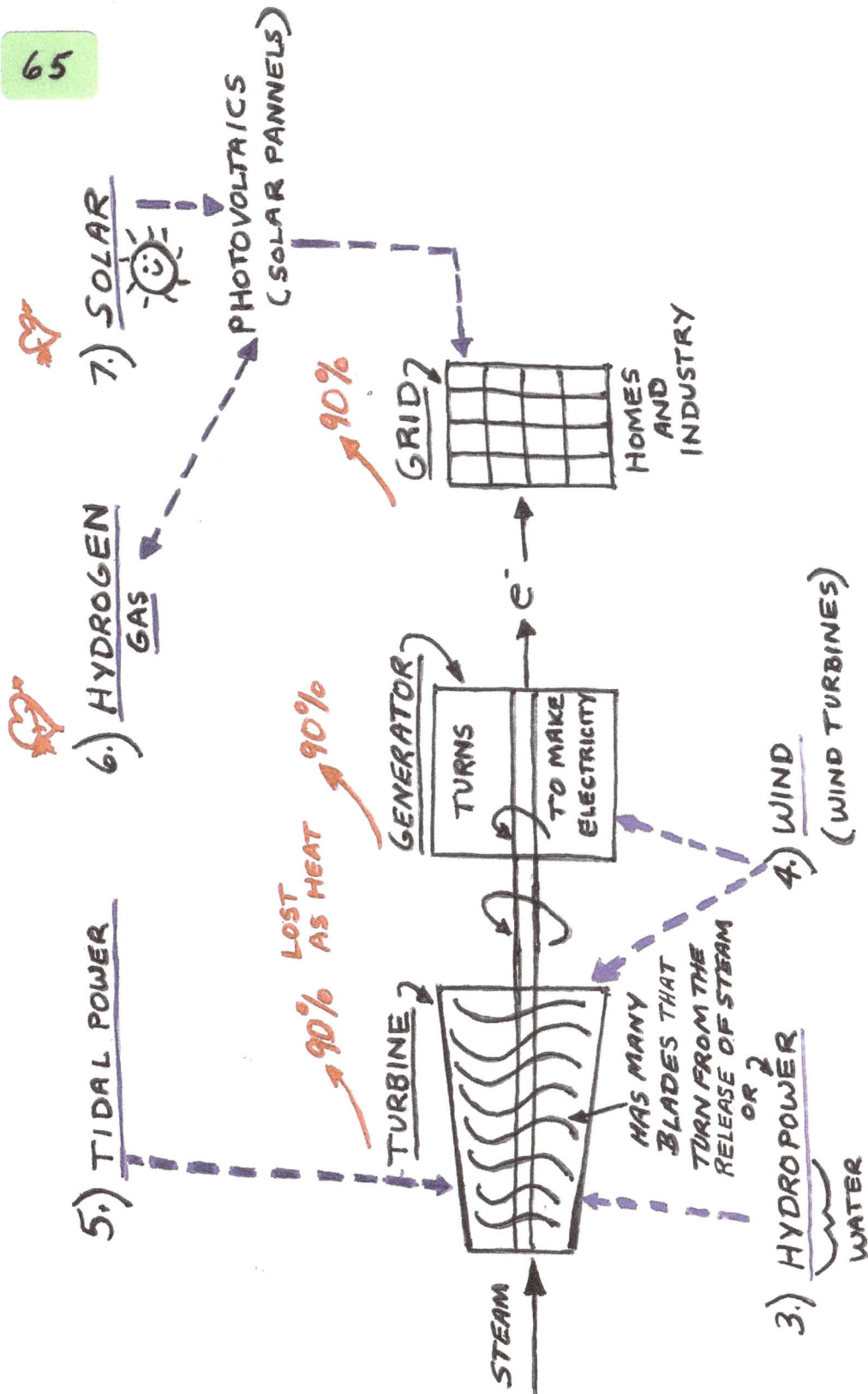

65

7.) SOLAR

PHOTOVOLTAICS
(SOLAR PANNELS)

6.) HYDROGEN
GAS

5.) TIDAL POWER

GRID

HOMES
AND
INDUSTRY

90%

e⁻

GENERATOR

TURNS

TO MAKE
ELECTRICITY

90%
LOST
AS HEAT

90%

TURBINE

HAS MANY
BLADES THAT
TURN FROM THE
RELEASE OF STEAM
OR ?

4.) WIND
(WIND TURBINES)

3.) HYDRO POWER
WATER

STEAM

Japan during World War II. (Notes 69, page 153 and Notes 70, page 154)

After much research and development, scientists found a way to control the nuclear reaction so the release of energy is not all at once. Remember the firecracker example? They learned how to slow the kaboom down and use the heat energy released to produce super-heated, high pressure steam. In this case, the nuclear process uses a series of control rods that limit the release of neutron energy. (Arms, 2008) This limit on the further breaking of other atoms slows, or controls, the release of heat energy that results from the reaction. This is also known as controlling the chain reaction. (Notes 69, page 153) This release of energy, from the breaking of the atoms and their nuclei, results in heat energy that turns the water into steam. The steam is then released into the turbine, and the turbine turns the generator, sending electrical power onto the grid for its use. While this process does produce enormous amounts of energy with little pollution, it was once considered the wave of the future as a major producer of electrical power. That optimism didn't last long once the downsides of using nuclear-based fuels became a reality. Using these facilities generated very toxic and dangerous wastes that had to be handled and stored safely. For example, the control rods and other equipment needed to be changed out on a regular basis to keep the reactor running, (Notes 70, page 154) not to mention the potential for nuclear accidents, such as in Chernobyl, Ukraine (1986). This accident rendered the whole city unsafe to live in. (Lallanilla, 2013, Sept. 25) In 1979, an incident at the Three Mile Island nuclear power plant in Pennsylvania almost resulted in a potentially dangerous radiation leak. While contained, it could have turned into a major environmental problem. (Three Mile Island, 2002, Jan.) Consequently, the planning and building of new nuclear reactors was put on hold because of the concerns for storing of radioactive hazardous wastes and the potential for accidents. Hey, a great song just came to mind It's called, ♪ *The Neutron Dance* ♪ by the Pointer Sisters (1983). It's pretty "funky" tune! Check it out! OK, keeping-er going!

Fossil fuels have their downsides also. Even if one could burn the fossil fuels completely, they would still be releasing carbon dioxide (CO_2), a gas that has a direct correlation with the greenhouse effect, contributing to the global warming problem. And, I didn't even mention all the coal ash that needs to be landfilled. Also, there is an efficiency factor with fossil fuels and nuclear fuels. Remember what the first and second laws of thermodynamics stated? We cannot make more energy; however, we are very good at converting or using it up, and every time it's converted, there is around a 90% loss of energy. I would like you to think about all the conversions that fossil fuels go through just to become the energy source we use to produce and place electrical energy on the grid. Take a look at the total conversions (Notes 64, page 144 and Notes 65, page 145) that occur in the production of electricity.

The photosynthesis process itself loses 90% of solar energy. The consumers, in order to grew and reproduce, lose another 90%. The decomposition and/or conversion of dead plants and animals to fossil fuels another 90% loss. Now we use the fossil fuels to make steam, another 90%, turn the turbine, another 90%. Turn the generator, another 90% loss. Then there is the grid, another 90% loss. WOW! That's seven conversions!

That's just an estimate, give or take the insulating and recovery systems being used, of the many conversions that occur with such a heavy loss of energy just to bring power into our homes. One needs to appreciate how far removed these processes are from the sun's original energy values.

So, what are the alternatives to nonrenewable energies? How can we keep up with the ever-growing demand for electrical power and not alter our climate or kill ourselves with pollution and/or radioactive wastes? How can we turn the turbines and generators without so many conversions? How can we better utilize the sun's energy more directly? In other words, we need to consider other energies that are stored or created by natural processes that supply us with energy. There is a lot of energy that can be harnessed from moving masses of air and water or, better yet, using the sun's energy directly. This is the premise behind using renewable energies.

Although there are many renewable energies being used and many new ideas being engineered, I'll look at seven renewable energies that are either the most popular or are gaining popularity in the production of electrical power. Again, remember to look at Notes 64, page 144, and Notes 65, on page 145, as I discuss these alternative energies.

The first renewable that would keep the fires stoked at the power plants is "Biomass." (Notes 71, page 155) This is the heat energy obtained from the burning of organic materials, or plant-based materials, that can be replanted. It can also be obtained from the products we already made, used, and discarded that originally came from plant tissues. For example, wood of all varieties, like branches, wood chips, and saw dust, are all sources of biomass energy.

By sorting out all the hazardous material, and with the right kind of air pollution control systems, our trash could become a source of biomass energy. If incinerated correctly, discarded furniture, tires, upholstery, etc. could be chipped, ground, and burned for their energy.

3 TYPES OF FOSSIL FUELS

* MADE UNDER THE CONDITIONS DISCRIBED EARLIER

1.) COAL (NON RENEWABLE)

COAL ADVANTAGES :)

- HIGH "E" VALUE
- CAN MAKE OTHER FUELS FROM IT

EX.) SYNFUELS
→ SYNTHETIC FUELS FROM COAL

DISADVANTAGES :(

- HEAVY ENV. COSTS

A. LAND → W/ STRIP MINING

B. AIR POLLUTION → BURNING, IT GIVES OFF CO_2, NO_x, SO_x

C. PRODUCES UNWANTED BYPRODUCTS → TAILINGS

D. HEALTH PROBLEMS
EX.) BLACK LUNG FROM COAL DUST

2.) PETROLEUM (OIL) (NON RENEWABLE) 67

- CREATED UNDER THE SAME PROCESSES AS COAL

- CONTAINS MORE ORGANIC SOLVENTS

Ex.)

CAN BE SEPARATED INTO MANY FUELS BY DISTILLATION PROCESSES

ADVANTAGES

- CHEAP TO EXTRACT AND TRANSPORT
- EASY TO PROCESS
- HAS A HIGH "E" OUTPUT

DISADVANTAGES — YIKES

- OIL SPILLS
- GROUND WATER CONTAMINATION
- RELEASES AIR POLLUTANTS
    Ex.) CO_2, NO_x, SO_x

* LIMITED SUPPLY

* USED MOSTLY FOR

- TRANSPORTATION
    LUBRICATION & FUELS
    Ex.) CARS, TRUCKS, JETS

- SOME AS HEATING FUEL

3.) **NATURAL GAS** (NONRENEWABLE) 68

- CREATED THE SAME AS COAL AND OIL

- NAT. GAS → RESULTS FROM DECOMPOSITION AND BUILDS UP UNDERGROUND

Ex.) MOSTLY METHANE

$$(CH_4)$$

ADVANTAGES THATS WORKS FOR ME

- LOTS OF RESERVES REMAIN
- BURNS CLEAN
- LOW COST

DISADVANTAGES

- DIFFICULT TO TRANSPORT "UNLIKE OIL → GAS IS VERY EXPLOSIVE"

USED MOSTLY FOR

- HEATING
- ELECTRICAL POWER PRODUCTION

Another form of biomass that's being harnessed is coming from landfills- - the production of methane (CH_4) gas. Remember, the anaerobic decomposers and their gases we discussed earlier? (Notes 42, page 95) landfills are biologically active with many any anaerobic bacteria. In the process of decomposing discarded organic materials buried in landfills, these anaerobic organisms produce methane gas just as it was produced in the formation of fossil fuels. When collected, it has a very high energy value. By burning the sources of biomass, its energy can be released to turn water into steam which turns the turbines, etc. In many cases, it's used as a supplemental fuel source to run the engines that turn generators that places electrical energy on the grid.

Biomass has great potential energy that could be harnessed only after we recycle as much paper, plastics, glass, and metals as possible. The organic-based materials that are unrecyclable but still hold energy values could be incinerated, while those items that cannot be burnt correctly would be landfilled as necessary. If we start thinking more along these lines, we could be making great strides to a more sustainable future. By seeing our discarded waste and trash as an energy resource, instead of just stuff that's filling our landfills, it creates a new way of thinking ecologically. As stated before, we live in a "Closed System." However, a word of caution should be noted, just like fossil fuels, even if these biomass materials are incinerated completely and correctly, they would still be releasing large amounts of carbon-dioxide (CO_2) into our atmosphere, adding to the global warming problem. The beauty of using biomass is that plants, given the right conditions and planning, can be replanted helping with our future energy needs. Plus, it has an added bonus of absorbing carbon-dioxide (CO_2).

These future plants help complete the cycling of carbon back into the tissues of more biomass, reducing the effects of carbon dioxide on climate change.

Geothermal is another renewable energy source (Notes 72, page 156). The steam produced from water that has been heated by volcanic action is pumped to the surface and used to turn the turbine, which turns the generator, which produces the electrical power placed on the grid. The condensed water is pumped back into the earth's crust to be reheated. While this energy is pollution free, it's not readily available everywhere. This source of energy is being utilized in areas of the world where the earth's crust is actively moving and volcanic magma moves closer to the earth's surface.

Another way to turn the turbines is by harnessing the energy contained in moving water. Here, the stored energy of water is held behind the walls of a dam. In turn, the water is channeled into spillways that flow over and through the turbine blades, and the turbine turns the generator that produces electrical power which is fed onto the grid. This power supply process is called "hydropower" or hydroelectric power (Notes 72, page 156). There are, of course, thousands of hydroelectric dams throughout the world. In my home state, the Wisconsin River has 25 hydroelectric power facilities. (Hydro plants, 2018). In the United States, the Hoover Dam is one of the largest, most famous hydroelectric power plants ever built. Every semester I would have some students who had the privilege of seeing it firsthand. It's on my bucket list. If time allowed, I would show my classes the American Experience documentary, "Making of the Hoover Dam." (Samels, 2006) What a great video that everyone needs to see!

Wind power has become a very popular form of renewable energy (Notes 73, page 157). In this case, the wind turns large propellers. These act as the turbines which turn the generator to place power right on the grid. The wind power, geothermal energy, and hydropower processes remove the added steps of heating water to steam to turn the turbines. This increases the efficiencies of electrical power production by reducing the amount of energy lost as heat from the burning of fossil fuels, plus it comes closer to using the sun's energy directly.

Tidal power, another form of hydropower, has been developed and is being used by some countries. (Notes 74, page 160) This is another renewable energy source that uses the energy contained in the movement of ocean tides created by the gravitational pull of the sun and moon on the Earth. (Bellis, 2018, March 19) The premise here is to take advantage of very predictable daily tides, or moving water. Designed around coastal areas, when the tides come in, the water is allowed to flow into a dammed-off area. Just before the tide goes out, the dam's gates close. This traps the ocean water. This water is then channeled to turbine gates that are opened.

As the water flows out, it turns the turbines which turns a generator that places power on the grid.

The next two renewable energies are very dear to my heart! And one wonders ♫ What Love Got to do With It... ♫ I especially love these next two!" Not only are these two energy sources versatile in their uses, but their processes are

69 | NUCLEAR FUEL (NONRENEWABLE)

USES "E" FROM THE BREAKING OF NUCLEI

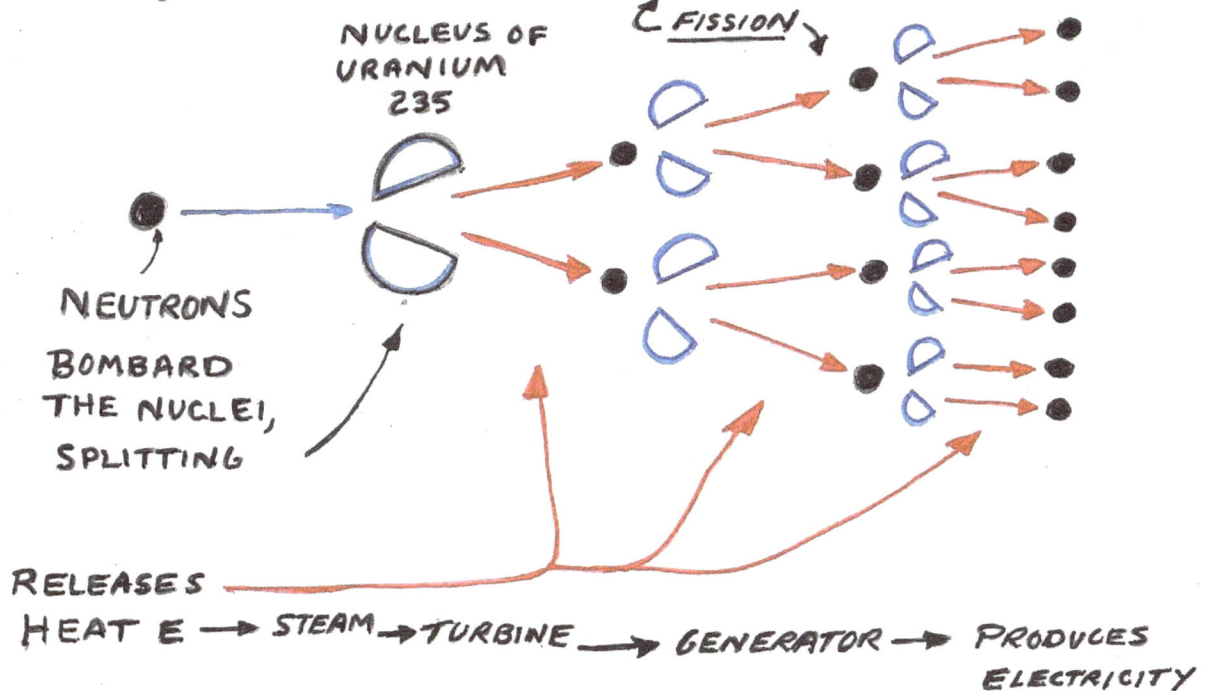

NUCLEUS OF
URANIUM
235

FISSION

NEUTRONS
BOMBARD
THE NUCLEI,
SPLITTING

RELEASES
HEAT E ⟶ STEAM ⟶ TURBINE ⟶ GENERATOR ⟶ PRODUCES
ELECTRICITY

ADVANTAGE 😊 — WOW!

- YIELDS ENORMOUS AMOUNTS OF "E"

1 Kg U = SAME BTU'S AS 2000 TONS
OF COAL

- LITTLE POLLUTION
MOSTLY EXCESS STEAM

DISADVANTAGES

Yuck!

- NUCLEAR ACCIDENTS

  Ex.) CHERNOBYL
  (UKRAINE 1986)

- WASTE DISPOSAL OF SPENT FUEL RODS AND EQUIPMENT USED IN THE REACTOR "HIGHLY RADIOACTIVE"

USED FOR

PRODUCTION OF ELECTRICITY (MOSTLY)

154

1.) **BIOMASS** (RENEWABLE)

- HEAT "E" FROM ORGANIC MATERIALS

- HEAT "E" FROM PLANTS

  Ex.) - WOOD

    - SAWDUST

    - BRANCHES

    - METHANE GAS
      OFF LANDFILLS

    - TRASH

  ALL CAN BE BURNED TO
  HEAT WATER AND TURN
    IT INTO

                ↓

        "STEAM"

TREES

JUST KEEP GROWING ME!

2.) **GEOTHERMAL**   (RENEWABLE)

LAND OR EARTH   HEAT

H_2O LIQUID → STEAM → TURN THE TURBINE → GENERATOR → PRODUCES ELECTRICITY

HOT ROCKS W/ PRESSURES

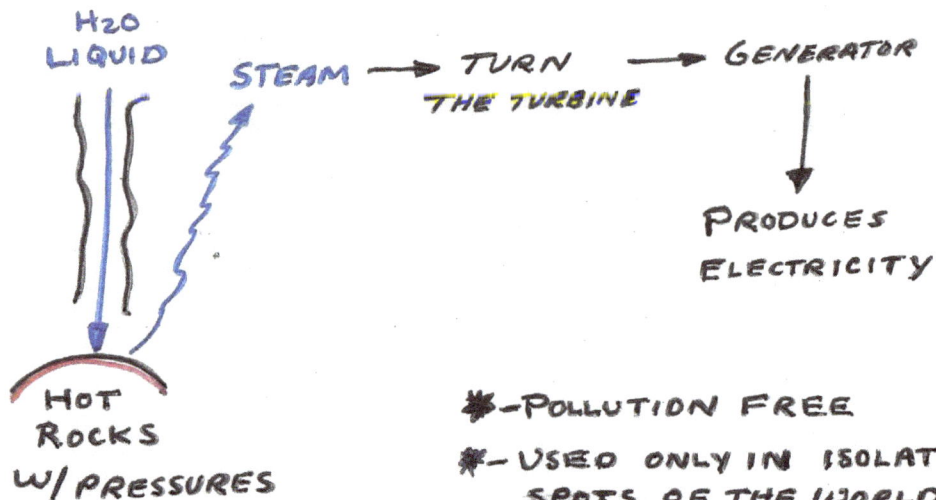

✱-POLLUTION FREE
✱-USED ONLY IN ISOLATED SPOTS OF THE WORLD W/ VOLCANIC ROCKS

3.) **HYDRO POWER**   (RENEWABLE)

WATER

FALLING/MOVING WATER TURNS THE BLADES OF THE TURBINES → TURNS THE GENERATOR → MAKES THE ELECTRICITY (HYDROELECTRICITY)

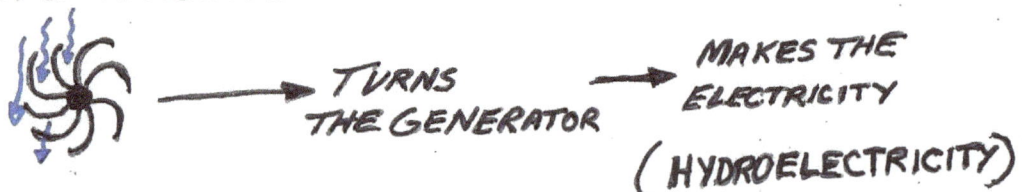

EX. HOOVER DAM

4.) **WIND POWER** (RENEWABLE)

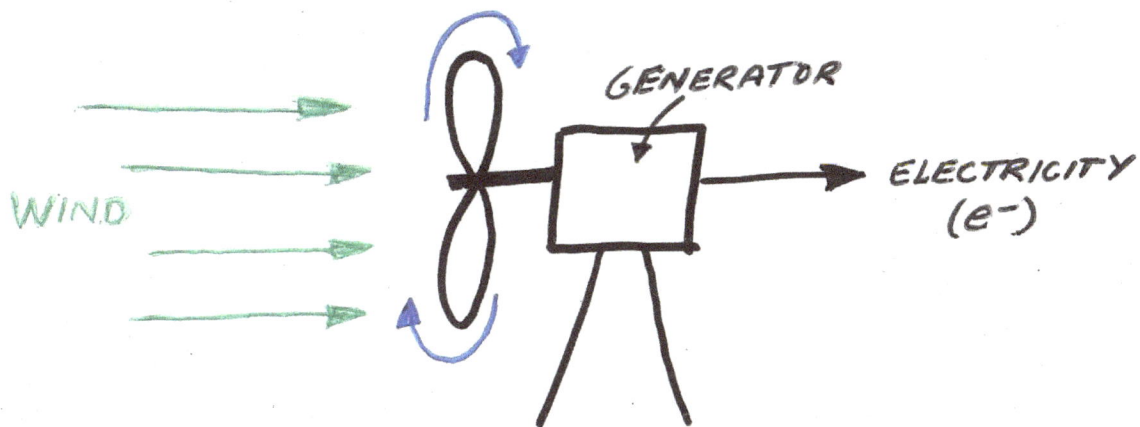

WIND

GENERATOR

ELECTRICITY
(e⁻)

WIND TURBINE

TURNS THE GENERATOR
DIRECTLY TO PRODUCE ELECTRICITY

157

the closest yet of all the renewables to using the direct energy of the sun. They are hydrogen and solar energy.

The beauty of using hydrogen gas as a renewable energy source is that it is contained in many of the substances that surround us in the biosphere. For example, hydrogen is an element found in all living things and is especially abundant in the form of water. What makes hydrogen gas such an excellent energy source is that it is very flammable and can be used directly for a variety of energy needs. As an example of its flammable characteristics, I would tell my students about the tragic Hindenburg disaster of 1937. The Hindenburg was a German Zeppelin that was full of hydrogen gas used to float it up into the air like a balloon. While docking in New Jersey, it caught fire and exploded. (Hindenburg, 2018) Because of its flammability, hydrogen gas can be used to directly run the internal combustion engines of cars and trucks, or it can be used as a fuel for heating homes, or turning water into steam to generate electrical power. Additionally, hydrogen gas can be used to generate electrical power without its combustion. This can be done by using a fuel cell technology that has been developed. Simply stated, this process begins by releasing the hydrogen gas through a membrane into the contents of these specialized metals. The hydrogen comes in contact with oxygen, creating a chemical reaction that causes the jumping of electrons resulting in the flow of electrical current. In turn, the electrical current can be used to run an electric car or power a home that is off the grid. Whether it's burned directly or released across a fuel cell, in either case, hydrogen reacts with oxygen creating water. As everyone knows, water is everywhere in the biosphere, making it essentially pollution free. That is why hydrogen, when extracted as a gas, is one of the most versatile sources of renewable energies.

The downside of hydrogen is in its extraction process. This is where the two laws of thermodynamics come into play, making the cost benefit unattractive with the technologies, energy infrastructure, and distribution systems in place today. While there are many methods of separating hydrogen gas from assorted organic and inorganic substances, most of the hydrogen gas produced is extracted from methane gas. Consequently, the energy requirements needed to separate it out offset the net amount of energy gained in the process. The most promising extraction process, and the one I love, is to extract hydrogen gas through an electrolysis process. (Rifkin, 2002) Currently, there are many process designs either being engineered or in use. In either case, they all use the sun's energy more directly and efficiently to produce electrical power. The electrolysis process best aligns itself with the first and second laws of thermodynamics. Electrolysis doesn't pretend to make energy, it just uses the renewable energy from the sun to transform, or split, other compounds to release their hydrogen contents. And, water is the perfect compound. To me, it's a match made in heaven, again, everyone wonders, "what's love got to do with it?" As I stated earlier, water not only surrounds us but holds a lot of hydrogen for the extraction process.

In the simplest of terms, the extraction process begins with the sun's energy striking the specialized metals in a solar panel. The sun's energy causes the atoms of these metals to begin jumping their electrons in a similar way that hydrogen did through the fuel cell to generate a flow of electrical current, only now, solar energy is being transformed into the generation of electrical current. The power generated from the solar panels is used to run the electrolysis process that releases the hydrogen atom from the water molecule. Electrolysis means electro, using electricity to lysis or split apart. (Notes 75, page 161) The best part of using solar panels is they can be used in close association with fuel cells to produce electrical power 24 hours a day, 7 days a week. Here's how. During the day, the solar panels are supplying enough electrical energy directly to meet the energy needs of a home. At the same time, the solar panels run an electrolysis system to produce and store hydrogen gas. In turn, the hydrogen gas is used in fuel cells at night providing a continuous supply of electrical current. Again, hydrogen has the potential to provide a variety of alternative methods of producing renewable energies to meet the requirements of a variety of different needs.

Now the question becomes, "So why not use solar energy directly?" We could, is the overwhelming answer. When large sections of solar panels are in place, electrical current is then fed right onto the grid, a process being used and gaining popularity around the world. Again, referring back to the two laws of thermodynamics, this process would cut out even more of the transformations and the subsequent loss of solar energy as heat. This would hold true even with all the other renewal energies I discussed. I would remind you the sun is shining somewhere all the time. All that is needed is a series of very large solar fields and a large enough grid to accommodate the earth's daily shifts from daylight to nighttime. (Notes 76, page 162)

5.) **TIDAL POWER** (RENEWABLE)

— USES COASTAL TIDES
THAT MOVE WATER IN
AND OUT WITH THE
GRAVITATIONAL PULL
OF THE MOON AND SUN ON
THE EARTH

✳ A FORM OF HYDRO POWER

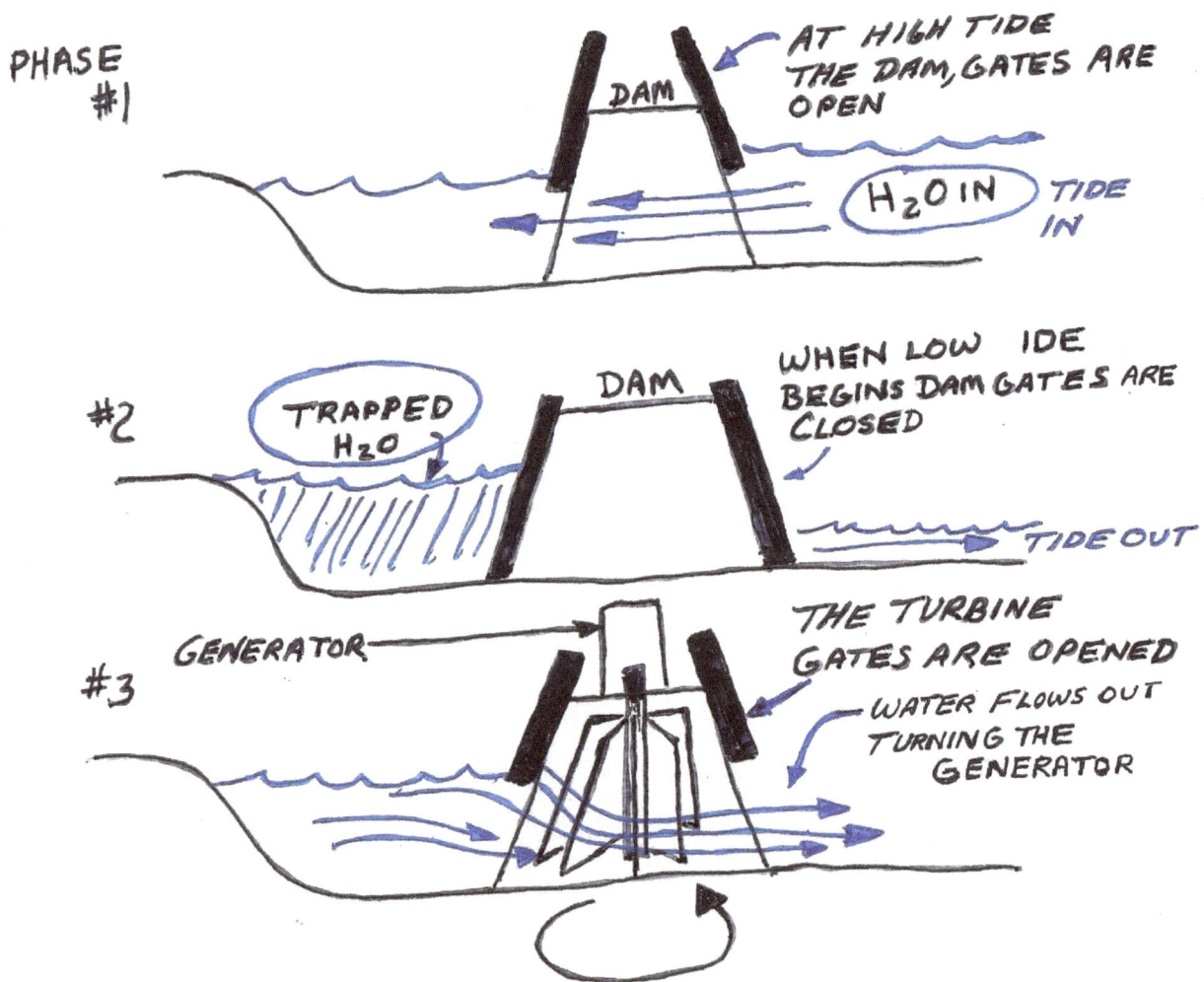

PHASE #1

DAM

AT HIGH TIDE
THE DAM, GATES ARE
OPEN

H_2O IN   TIDE IN

#2

TRAPPED H_2O

DAM

WHEN LOW IDE
BEGINS DAM GATES ARE
CLOSED

TIDE OUT

#3

GENERATOR

THE TURBINE
GATES ARE OPENED

WATER FLOWS OUT
TURNING THE
GENERATOR

REMEMBER THE HINDENBURG

6.) **HYDROGEN (H_2)** (RENEWABLE)

USES PHOTOVOLTAICS OR SOLAR NELS FOR ELECTRICAL "E"

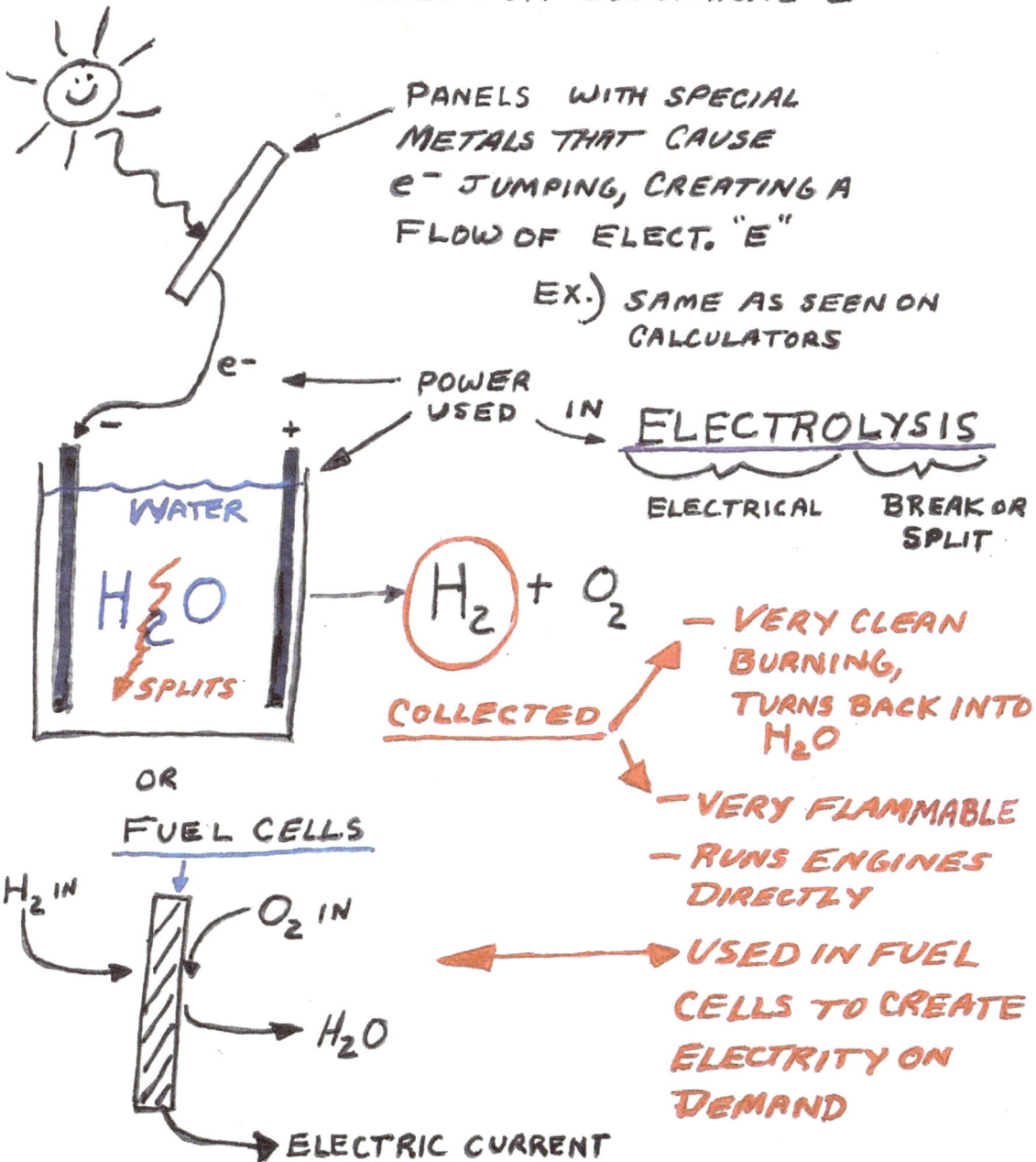

PANELS WITH SPECIAL METALS THAT CAUSE e^- JUMPING, CREATING A FLOW OF ELECT. "E"

EX.) SAME AS SEEN ON CALCULATORS

e^-

POWER USED IN **ELECTROLYSIS**

ELECTRICAL   BREAK OR SPLIT

WATER

H_2O SPLITS

$H_2 + O_2$

COLLECTED

- VERY CLEAN BURNING, TURNS BACK INTO H_2O
- VERY FLAMMABLE
- RUNS ENGINES DIRECTLY

USED IN FUEL CELLS TO CREATE ELECTRITY ON DEMAND

OR

FUEL CELLS

H_2 IN

O_2 IN

H_2O

ELECTRIC CURRENT

7.) [SOLAR] (RENEWABLE)

↳ USING THE SUN'S ENERGY DIRECTLY TO GENERATE ELECTRICITY

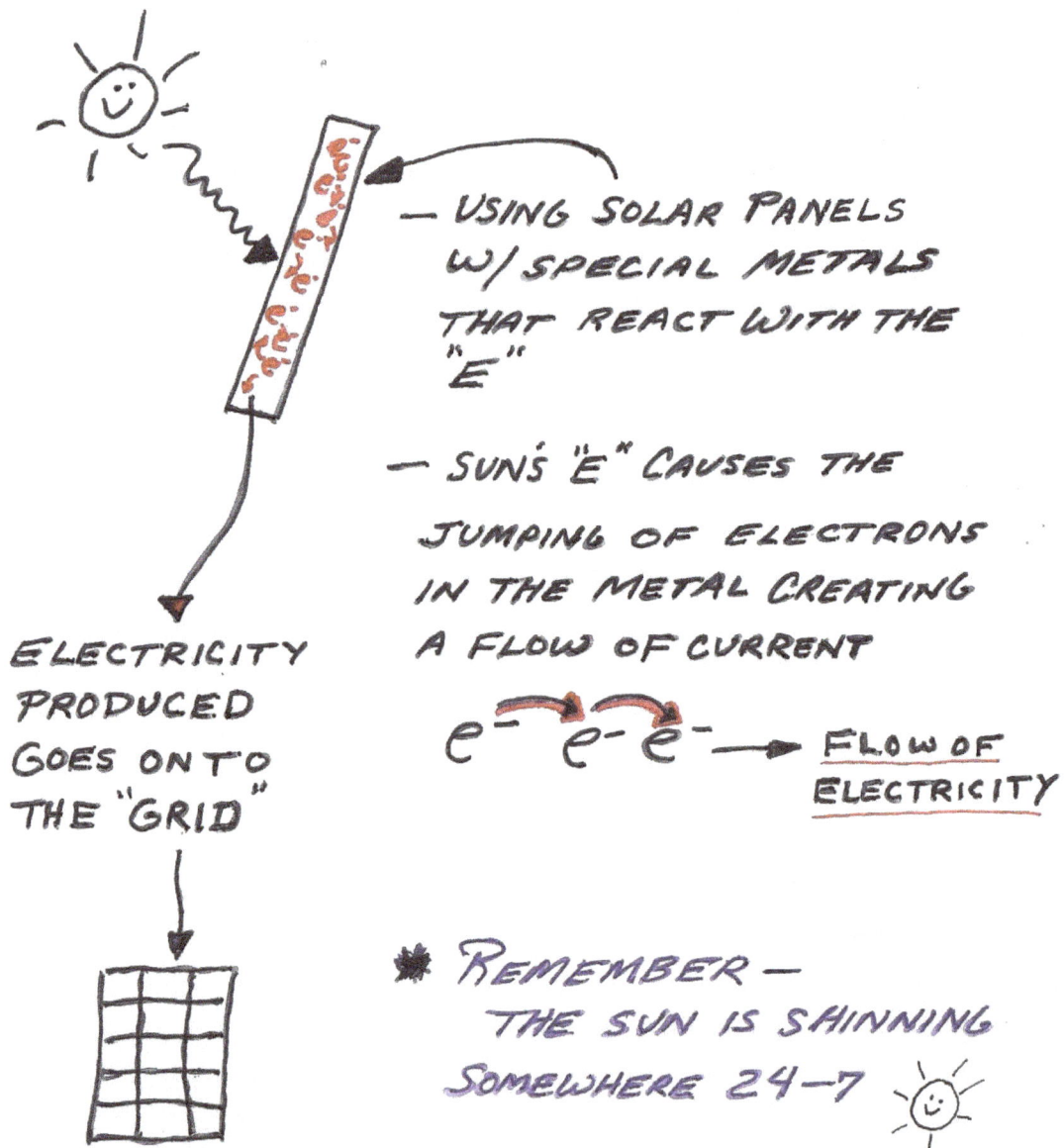

— USING SOLAR PANELS W/ SPECIAL METALS THAT REACT WITH THE "E"

— SUN'S "E" CAUSES THE JUMPING OF ELECTRONS IN THE METAL CREATING A FLOW OF CURRENT

e^- e^- e^- → FLOW OF ELECTRICITY

ELECTRICITY PRODUCED GOES ONTO THE "GRID"

✱ REMEMBER — THE SUN IS SHINNING SOMEWHERE 24–7

162

I believe diversity of energy production is the way of the future. We need to stop thinking along the lines of an all or nothing concept. Such has been the case of fossil fuels. By slowly reducing our dependence on fossil fuel-based energies and substituting them with more sustainable renewable energies, is a policy needed for our future. By doing so, it will reduce our carbon emissions while still meeting the ever-growing demands for electrical power. Diversifying will be a critical step towards heading off or controlling global warming problems. Remember the carbon cycle, the law of conservation of mass, and the fact we live in a closed system - - all have to be considered as we move forward. Needless to say, the way we produce and use energy is critical to our survival. To finish up this section I would like everyone to remember what Chief Seattle said, "We are merely strands in the web of life. . .what we do to the web, we do to ourselves." (1854)

It's tune time ♫ *Electric Ave* ♫ by Eddie Grant (1982). This is a "fab" song you need to check out!

SECTION FOUR

A Lesson on Who's Eating Who and Why

(Can you find the "K2?")

78

REMEMBER THE 3 REQUIREMENTS
FOR ALL ECOSYSTEMS

1.) <u>NUTRIENT CYCLING</u> ✓

2.) <u>ENERGY FLOW</u> ✓

3.) <u>STRUCTURE</u>

3.) ECOSYSTEM STRUCTURE

OR

PREDATOR-PREY RELATIONSHIPS

* REVIEW

STRUCTURE   VS.   FUNCTION

THE RELATIONSHIPS
BETWEEN ORGANISMS
(<u>BIOTIC</u>) AND THEIR
NON LIVING (<u>ABIOTIC</u>)
ENVIRONMENTS

ALLOWS LIFE TO
EXIST

CHIEF SEATTLE (1854)
" WE ARE ALL CONNECTED..."

As I did with previous sections, I'll begin this section on ecosystem structure, also known as predator-prey relationships, with a short review of the scenario about the oak tree and its acorns. The oak tree obtains its nutrients from the biosphere, the land, atmosphere, and water, while obtaining its energy from the sun. When the squirrel eats the acorn and the hawk eats the squirrel, they do so in order to meet their nutritional and energy values for growth and reproduction. Through these actions they set up and maintain the third requirement of all ecosystems, its structure. Ecosystem structure is all the interactions between the abiotic and biotic portions of all ecosystems. (Notes 78, page 167) These interactions are also known as an ecological concept called "Predator-Prey Relationships." In fact, the very act of catching and consuming prey meets all three of the necessary requirements for all ecosystems to function and which allows life to exist. Consequently, in this section, I hope to present a clearer picture of nature's design behind predator-prey relationships by bringing together the information and laws covered in the previous sections. When studying predator-prey relationships, we need to return to a familiar concept called "Food Chain." This is one of the most common and earliest science topics taught at the grade school and/or middle school level.

FOOD CHAINS = PREDATOR - PREY RELATIONSHIPS

PRODUCER ⟶ 1° CONSUMER ⟶ 2° CONSUMER . . . .
　　　　　　　 (HERBIVORE)　　　　(CARNIVORE)

PLANTZ ⟶ GRASSHOPPER ⟶ BIRD ⟶ CAT

✱ NOTE: EACH FOOD CHAIN ENDS W/A
　　　　　　TOP CARNIVORE

NICHE — AN ORGANISMS
　　　　　　FUNCTIONAL ROLE (OR WAY OF
　　　　　　LIFE) IN A PORTION OF AN
　　　　　　ECOSYSTEM

EX.) NICHE OF GREEN PLANTS ⟶ PRODUCERS

FOOD WEBS
　　　　— A SERIES OF FOOD CHAINS
　　　　　THAT ARE CROSS - LINKED

　　　　— NOT ALL ORGANISMS ARE ASSIGNED
　　　　　TO ONE TROPHIC LEVEL

　　　　　EX.) HUMANS (OMNIVORES)
　　　　　　　WE EAT BOTH PLANTS AND
　　　　　ANIMALS

A SIMPLE FOOD WEB

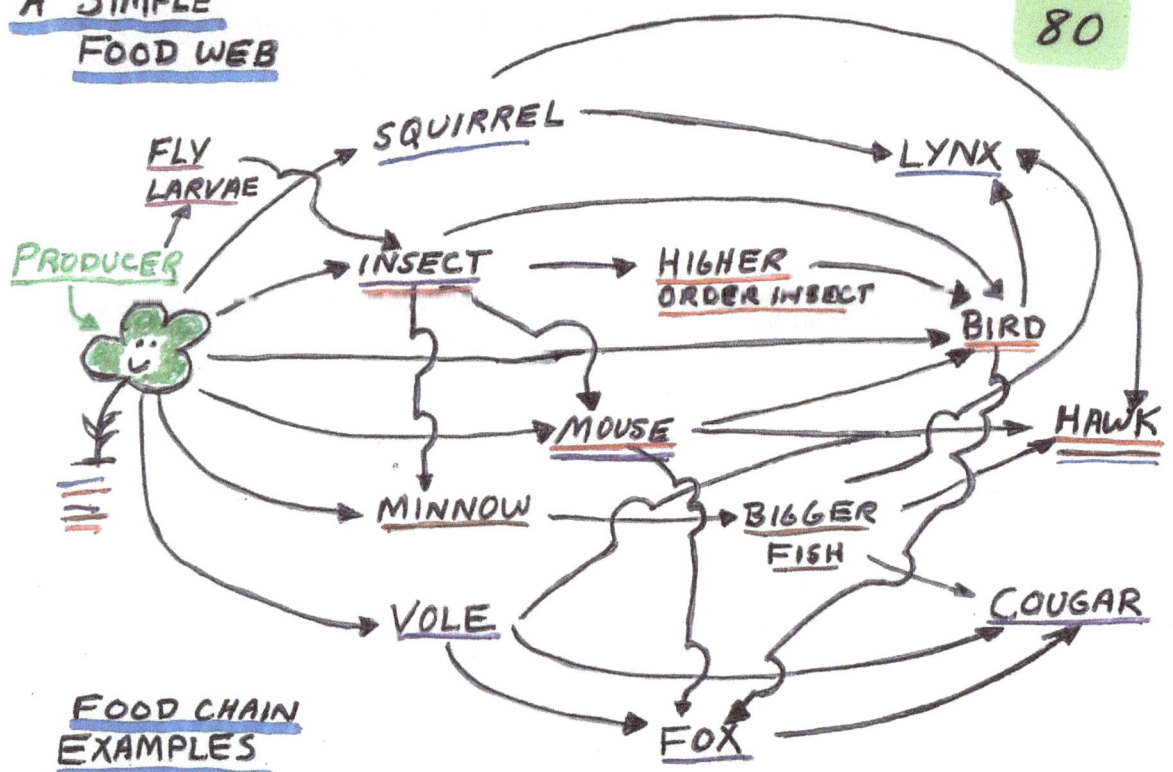

FLY LARVAE

PRODUCER

SQUIRREL

INSECT

HIGHER ORDER INSECT

MOUSE

MINNOW

BIGGER FISH

VOLE

LYNX

BIRD

HAWK

COUGAR

FOX

FOOD CHAIN EXAMPLES

PRODUCER	1° CONS.	2° CONS.	3° CONS.	4° CONS.
ACORNS	SQUIRREL	HAWK		
GRAIN	MOUSE	FOX		
GRAIN	BEETLE	MOUSE	FOX	
GRAIN	FLY LARVAE	BEETLE	MOUSE	FOX
"	"	"	"	"
"	"	"	"	"
"	"	"	"	"
"	"	"	"	"
"	"	"	"	"

Now with the background information I've provided I hope to give you a broader interpretation of the term "food chain." (Notes 79, page 169) In its simplest definition food chains are nothing more than predator-prey relationships. (Notes 79, page 169) We already covered why critters eat and/or are eaten, which is for the nutritional and energy values needed for survival. However, when food chains become interconnected, they move to a higher level of organization known as a "Food Webs." (Notes 80, page 170) Food webs are set up by a concept called "Trophic Levels," or feeding levels. These trophic levels are very similar to the order of events with the nutrient cycles. (Notes 81, page 172) It always starts out with the producers. However, when a food web develops, multiple trophic levels are possible. This occurs because of the variety of food choices available depending on who's eating who. In other words, if a food web is broken down into all of its possible food chains, one critter might have a different position of feeding or being feed upon. (Notes 80, page 170 and Notes 81, page 172) With all the variable trophic levels that occur in various ecosystems, researchers developed a numbering system for each of the consumers as they relate to their individual food chains. As usual, it starts with the producers. However, as I outlined earlier, consumers vary in their feeding habits. Some are herbivories that feed directly on plant tissues, some are omnivores that eat both plant and animal tissues, and others are carnivores who strictly eat meat.

PRODUCERS   EX.) PLANTZ

↓

PRIMARY CONSUMER (1°)   EX.) HERBIVORS
- PLANT EATERS ONLY
EX.) DEER, COWS, INSECTS

↓

SECONDARY CONSUMER (2°)   EX.) CARNIVORES
- MEAT EATERS ONLY
EX.) - CATS
- FOXES
- INSECT EATING BIRDS

↓

TERTIARY CONSUMER (3°)   OMNIVORES
- EAT BOTH PLANTS AND ANIMALS
EX.) - HUMANS
- BIRDS

↓

QUATERNARY CONSUMER (4°)

DECOMPOSERS (MOSTLY BACTERIA AND FUNGI)
CONSUMERS THAT GET THEIR FOOD BY BREAKING DOWN DEAD ORGANISMS → CAUSING ROT

This is why consumers had to be divided into trophic levels. The trophic levels are ranked as Primary (1), Secondary (2°), or Tertiary (3°) consumers. Sometimes there is a fourth level called Quaternary (4°). (Notes 81, page 172) This depends on the ecosystem and its biodiversity. So, "let the food chain games begin!" I would like you to look at my simple food web. (Notes 80, page 170) In turn, I would like you to focus on the various food chains that make it a web. Look at notes 80 (page 170) and notice the food chains I've provided. The mouse could be a primary (1°) consumer by eating only the grain. However, if the mouse eats the insect that ate the grain, the mouse becomes the secondary (2°) consumer. To demonstrate further, if the fly larvae eats the grain, it becomes the primary (1°) consumer. And if an insect, perhaps a beetle, feeds on the larvae, the beetle becomes the secondary (2°) consumer. Finally, should the mouse feed on the beetle, the mouse would become the tertiary (3°) consumer. Notice what happens to the fox's trophic level when the food chain changes. (Notes 80, page 170) This is a simplified example, but I think it demonstrates the need for designated trophic levels depending on the food chains and the number of individual species that make up a particular food web.

It's important to know how individual trophic levels are labeled and arranged because they dictate another shape, or pattern, that develops in all ecosystems. This pattern has become known as an "Ecological Pyramid." (Notes 82, page 174) At first glance, it seems very predictable that producers are at the bottom of the pyramid, creating the base of the pyramid. (Notes 82, page 174) When studying the food chain, it would seem the individual numbers of each trophic level would decrease while the sizes, or biomasses, of the individuals would increase. This may seem obvious to some but not to everyone. Scientists studying trophic levels started asking the question, "Does this pyramid shape remain constant with all ecosystems as individual food chains develop? If so, what drives this this naturally-occurring pyramid shape?"

ECOSYSTEM STRUCTURE
OR
PREDATOR — PREY RELATIONSHIPS

↓

FOOD CHAINS

↓

FOOD WEBS

↓

FOOD PYRAMIDS

TOP CARNIVORE

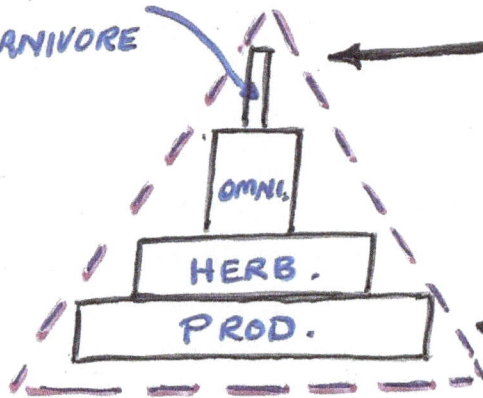

AT THE "TOP"
LARGE INDIVIDUALS
W/ SMALL POPULATIONS

OMNI.

HERB.

PROD.

AT THE BOTTOM
SMALLER INDIVIDUALS
W/ LARGER POPULATIONS

FOOD PYRAMIDS
SHOW THE TROPHIC
FEEDING LEVELS
OF THE FOOD CHAINS

BUT WHY THE SHAPE?

?

IS IT LOVE?

I would ask you to re-examine the simple food web and look closely at some of the food chains. (Notes 80, page 170) Do you notice any patterns developing with the different tropic levels?" At first observation it would seem straight forward. Most would assume it's only natural that as the number of individuals goes down their size increases, and this would be because bigger predators have a better chance to catch their prey and feed at each level. However, we cannot assume at face value the pyramid shape naturally occurs. It would have to be definitively proven. Consequently, ecologists had to take into account three possibilities for the pyramid shape. (Notes 83, page 177) The first was the "Pyramid of Numbers." This pyramid's shape was based strictly on the number of individuals of a species in a given ecosystem. However, this idea was quickly eliminated because researchers had documented that some food chains, in a variety of instances, would become inverted when using the number of individuals. (Notes 83, page 177 and Notes 84, page 178) And everyone knows what happens to a pyramid when it's placed on its peak. It falls over. Case in point. Ecologists found a parasitic organism where numbers actually increased as the trophic levels went up. For instance, ecologists found a group of birds that feed on a small population of fruit trees (or producers). By doing so, the birds became the primary (1°) consumer. But the birds also had parasites feeding on them, which would not be unusual, however, when examined closely and their numbers counted, it became apparent the parasites were greater in number than the birds on which they fed. This would make the parasites the secondary (2°) consumer. What was even more enlightening was a second parasitic population feeding on the first. Yes, even parasites have parasites! In this case, they found the count of the second parasitic species was higher than that of the original parasites being counted. As a result, the second parasitic species, with its higher numbers became the tertiary (3°) consumer in the food chain. In this case, the pyramid of numbers would become inverted. Consequently, species' numbers weren't responsible for the consistent pyramid shape. (Ecological Pyramids, n.d.)

Next, researchers looked at the size of the animal (or an individual's total biomass), that could possibly dictate the pyramid shape. (Notes 85, page 181) This is known as the "Pyramid of Biomass." It was so named because as individuals moved up the food chain, they became larger with less individuals per trophic level. This pyramid concept reasoned that individuals who are larger need more resources to survive. For example, larger animals would need more food, water, and land in order to meet their nutritional and energy requirements. Consequently, it was in the best interests of the species to limit its population to alleviate competition among themselves while trying to survive in an already dangerous predator-prey environment. Again, just like "pyramid of numbers," exceptions were also found with the "pyramid of biomass." In documented cases the pyramid of biomass was also found to be inverted. One such case involved researchers working in the English Channel who found zooplankton (consumers) in higher dry mass than the algae (producers) they were feeding on. (Campbell, 1996)

At first glance, the pyramids of numbers and biomass, or even a combination of the two, would undoubtedly solve the question of why the pyramid shape, but it didn't. The two didn't provide a scenario that would cover all the possibilities for the pyramid shape. Once again, I would like to refer you back to the prior notes on the simple food web and some of the food chains it produced. (Notes 80, page 170)

I would point out; all of the food chains were only to the third (3°) and forth (4°) trophic levels. Why? Because of the "Second Law of Thermodynamics." With about 90% of the energy being lost as heat within each trophic level, the energy runs out.

Consequently, except for a few unusual, or very diversified food chains, the third (3°) or forth (4°) trophic levels are usually the highest and most common levels that can be reached. In this case, Energy (E) flow becomes the limiting factor. (Remember Notes 50, page 112) As the result, the flow of energy regulates the length and scope of the food chains.

83 NOTE: W/ PREDATOR-PREY RELATIONSHIPS

MOVING UP

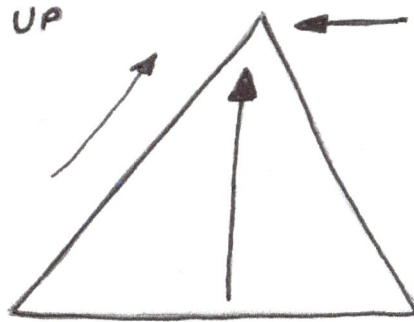

← LARGE INDIVIDUALS IN SMALL POPULATIONS

← SMALL INDIVIDUALS IN LARGE POPULATIONS

WHY?

AT FIRST GLANCE

THERE ARE

3 WAYS ECOLOGICAL PYRAMIDS DEVELOPED

1.) PYRAMID OF NUMBERS

2.) PYRAMID OF BIOMASS

3.) PYRAMID OF ENERGY

JUST LOOK !!

1.) PYRAMID OF NUMBERS

- AS THE SIZE OF THE ORGANISM INCREASES, THE NUMBERS OF THE INDIVIDUALS DECREASE FOR EACH TROPHIC LEVEL

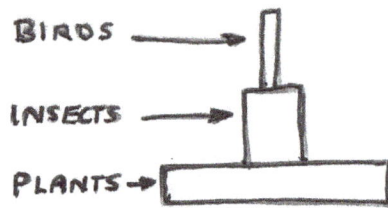

BIRDS →
INSECTS →
PLANTS →

DISADVANTAGE HERE:

ARE YOU KIDDING !?

- THIS IS NOT ALWAYS THE CASE

- IT HAS BEEN DOCUMENTED THAT THIS PYRAMID CAN BE INVERTED IN SOME CASES

EX. PARASITIC FOOD CHAIN

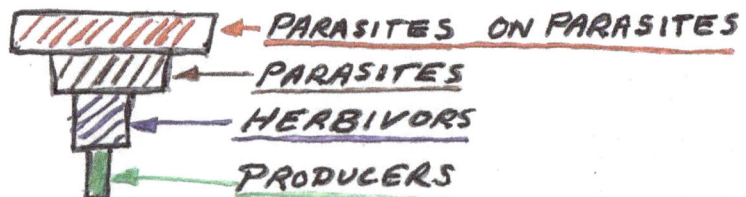

← PARASITES ON PARASITES
← PARASITES
← HERBIVORS
← PRODUCERS

This is further explained by the "Pyramid of Energy." When researchers began to test and calculate the energy flow for each trophic level from various food chains, the pyramid shape was consistently reproduced. When energy flow became the main focus of the pyramid shape problems with the pyramids of numbers and biomass were discovered. Both of these pyramids implied the energy values of tissue mass were equal per trophic level. This explained the inconsistencies. In other words, the first two pyramids were based on the assumption energy values of one gram of grass had the same energy value as one gram of grasshopper. (Notes 86, page 182) This is not the case.

I'm sure you have that very confused look on their face – the same kind as Abbott and Costello "who's on first…" look. Yes, ecologists have a way to compare the energy values in grass to those found in a grasshopper. For example, how can I compare the energy value of an apple to a candy bar? I can eat them both. What would give me more energy value, a delicious cheeseburger, which I love to eat, (notice the love), or should I go for the salad?" (Notes 86, page 182) If you were thinking calories, you're right. A popular way to watch ones diet these days. But what is a calorie? By definition, a calorie is the amount of heat energy needed to raise one gram of water one degree Celsius. (Petrucci, 1989) It's important to understand this calorie definition because it provides a way to compare energy values of apples to candy bars or grass to grasshoppers. The comparison of energy contained in the tissues of living things explains the consistent pyramid shape of food chains in all ecosystems, unlike the pyramids of numbers and biomass.

Here's how these calorie comparisons are calculated, and I'll keep it in the simplest of terms. However, I feel it's an important technique to discuss and most people find it interesting especially those dieting. The procedure for finding and comparing calorie contents of any organic-based materials begins by the assumption these materials can be burned. With that being the case, scientists developed an instrument called a "Bomb Calorimeter," which measures the change in temperature of heat energy released during the combustion process of organic materials.

The Bomb Calorimeter's design has a small combustion chamber surrounded by water that is monitored with a thermometer for changes in temperature. Inside the combustion chamber, equal amounts of dried materials that are being compared are burned individually. The burning, or combustion, releases the stored energy in each of the samples. In turn, the released heat energy increases a known volume of water that surrounds the combustion chamber. Using a few mathematical equations and the changes in the internal temperature, the calories can be determined. (Petrucci, 1989) This information allows ecologists to compare the energy values of each trophic level. They repeated these calorie tests on multiple trophic levels in a variety of food chains, and they all produced the repeating pyramid shape.

The incredible part about the pyramid of energy is it not only explains the pyramid's shape but it simultaneously answers the pyramid of numbers and biomass problems. Remember, the energy values explain the pyramids shape because of the laws of thermodynamics. With each conversion there is about a 90% loss of energy. (Notes 87, page 185) Again, this leads to less and less available energy for each individual trophic level, and results in populations having to eat more as they move up the food chain. With less available food, species' populations drop in numbers.

Plus, less availability of food creates more competition for resources, which results in larger individuals having the advantage of catching their prey. Hence, larger animals are more inclined to eat larger animals, or larger amounts, to meet their energy needs.

2.) PYRAMID of BIOMASS

- TAKES INTO ACCOUNT THE
TOTAL DRY MASS OF
EACH ORGANISM AT EACH
TROPHIC LEVEL

TOTAL DRY
MASS WOULD
DECREASE

DISADVANTAGE:

LOOK - - -→ THIS IS NOT ALWAYS THE CASE

✷ IT'S BEEN DOCUMENTED THAT
THERE ARE ECOSYSTEMS WHERE
THE PYRAMID IS INVERTED
WHEN CALCULATING TOTAL DRY
MASS

OH
OH

EX. ENGLISH CHANNEL STUDY

ZOOPLANKTON ← CONSUMERS

ALGAE ← PRODUCERS

3.) PYRAMID OF ENERGY

— EACH TROPHIC LEVEL
REPRESENTS THE TOTAL
"E" FLOW

THE OTHER PYRAMIDS (NUMBERS/MASS)

— IMPLIES THAT THE NUTRITIONAL VALUES
OF MASS ARE EQUAL PER TROPHIC LEVEL
"NOT TRUE"

"E" IN 1g OF FROSTED FLAKES $\neq$ "E" OF 1g OF BEEF

"E" IN 1g OF GRASS $\neq$ 1g GRASSHOPPER "E"

REVIEW OF CALORIES "PLEASE MAKE NOTE"

"E" VALUES $\longrightarrow$ FOR FOODS ARE IN CALORIES (cal)

CALORIE — IS THE AMOUNT OF "E" NEEDED TO RAISE
ONE GRAM OF WATER ONE DEGREE CELSIUS

$$1000 \, cal = 1 \, kcal = C$$
SEEN ON FOOD LABELS

Being the muscle man that I am (not!), which should I eat to give me the most sustaining energy source? The cheeseburger, or the salad? Of course, the cheeseburger – "It's so choice." And this is why the hawk (a raptor or bird of prey) prefers to eat the squirrel instead of cracking the nut for its survival. It generally comes down to evolutionary traits that were best suited for those individuals to obtain their nutritional and energy values without being eaten themselves. This is where Charles Darwin's Theory of Evolution explains natural selection and the survival of the fittest, which reinforced the design of the pyramid of energy. Those consumers that have the best suited traits, whether that be speed, color, or size, etc., will have the advantage or disadvantage of catching, or being caught, as a food source. In either case, the better suited an individual's traits are for a particular environment, the better their chances of passing on those traits and surviving as a species. (Campbell, 1996) If passing on traits is the key to a species survival, then in the case of larger individuals it becomes advantageous to the species to have smaller populations. Larger critters need more resources to obtain the energy requirements necessary to sustain their size, growth, and reproduction. Consequently, why expend the energy on having too many offspring that only adds to the competition/demand for resource availability between individuals of the same species and other large predators? In this case, survival of the fittest favored the traits of larger individuals who routinely produced less offspring for those reasons. Less offspring meant populations could sustain themselves without the tremendous loss of energy to produce more individuals only to be weakened and/or lost to starvation by competition among their own species, not to mention other large predators. (Notes 88, page 188) As a result, larger individuals tend to produce less offspring per year. This produces trophic levels that have larger individuals with less numbers in their respective populations.

The pyramid of energy, like the other pyramids, begins with a large base of producers. It's because the producers, remarkable in their varieties and numbers, capture most of the direct energy from the sun. Think about all the Oak trees and the acorns they produce, not to mention all other plants that produce edible fruits, seeds, roots, etc., that store and provide energy for other critters. Think about most of the primary (1°) consumers, such as mice, mole, voles, and other rodents that feed mostly on the producers. These guys have five to ten litters per year, with an average of six to eight, or more individuals per litter. (Bunker, 2018, Nov.18) Again, smaller individuals with higher populations. Another example is the fox, which feeds on plants and animal tissues. Being a larger animal and an omnivore, this places the fox on the primary (1°) or secondary (2°) trophic level. As the result, the fox breeds once a year with an average of five pups. (Fox, 2007) Again, larger individuals having even less offspring, for the same reasons I highlighted earlier, all regulated by the flow of energy. (Notes 89, page 189)

Many people ask me about the human population and the impact it's having on the natural pyramid shape. (Notes 90, page 190) I would pull out one of my quotes from "Get Smart" and say, "Chief, would you believe…" we are not on the top of the food chain. Think about surviving in the wild or swimming in the ocean. Between lions, tigers, bears (oh my) as well as sharks and killer whales, I can assure you we are not at the top of all food chains. Most people would quickly respond with the question about "bows and arrows" as well as "guns." I would respond with, "what about fossil fuels?" This is where they would give me the "Houston we have a problem…" look. I would remind them to think about the world's population right now, which is about 7 billion. (World Population, 2018) I'm not so worried about being eaten as I am about being able to produce enough food to feed that many people. The U.S. alone has about 321 million people.

NOW?
"E" VALUES OF
CAL. (OR Kcal.)
REPRESENT EACH
TROPHIC LEVEL

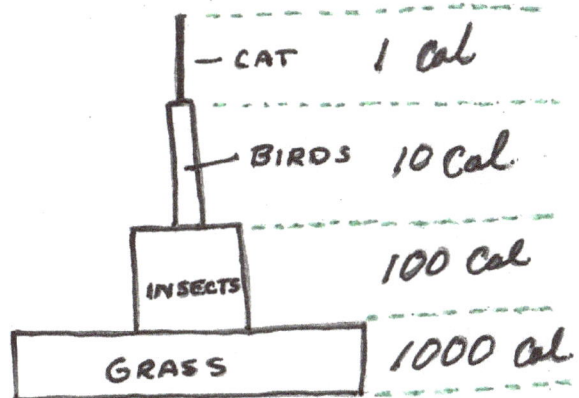

CAT   1 Cal

BIRDS   10 Cal.

INSECTS   100 Cal

GRASS   1000 Cal.

SHOWS A TRUE ORDER OF
FEEDING LEVELS

REMEMBER: 2ND LAW OF THERMODYNAMICS

FOR EACH TROPHIC LEVEL THERE
IS LESS "E" AVAILABLE TO DO WORK
CREATING THE PYRAMID SHAPE

What's really amazing is, even with these high populations, we can still produce enough available food, and then some. How can this be? The human populations have grown in significant numbers to invert the natural pyramid shape. (Notes 90, page 190 and Notes 91, page 191) Our populations have made the natural pyramid top heavy, and the First and Second Laws of Thermodynamics are still in place. Again, I ask, "How can this be?" The pyramid of energy says with each trophic level there is less and less available energy. If humans are the omnivores in the pyramid, depending on the food chain they are placed in, they could be primary (1°) or secondary (2°) consumers. We cannot defy the Laws of Thermodynamics. So, why hasn't the pyramid collapsed on us? Where is the energy coming from that props up the human-inverted shape? Yes, it's fossil fuels. As I said earlier, fossil fuels have made it easy for human populations to grow since the industrial revolution. (Notes 90, page 190 and Notes 91, page 191)

The increased mechanization of farming and food processing has led to increased production and availability of food to feed the ever-growing populations of the industrialized world. However, this is where the First Law of Thermodynamics becomes apparent. If humans cannot make more energy and can only convert it, then how do we work the trick of feeding increased populations? The answer was found in fossil fuels. As I demonstrated earlier, all our energy can be traced directly or indirectly back to the sun. The same can be done in this case, just trace the energy sources that helped sustain the increased human population for over one hundred fifty years. To illustrate how the laws of energy are still in control, we just need to follow the energy source that has propped up the imbalance of human populations that have essentially inverted the natural pyramid shape. It always comes back to fossil fuels. It also shows how dependent we are on them to produce and supply our populations with everyday needs. Remember, it always comes back to fossil fuels and the technologies that use them.

To give you an example of this I would like tell another story from my classroom days to prove my point. I would ask my students to give me an example where fossil fuels are not directly or indirectly involved in their everyday lives. I would say, "Go ahead, make my day." Knowing what they have already learned, my students would hesitate. I remember a witty soul answered, "The water we drink." My reply, "Are you referring to the water we drink, wash our bodies and clothes with, and that carries away our sewage? That's easy, all the water we use is delivered to us by fossil fuels." In the Milwaukee area, water is first pumped from a freshwater source, in this case Lake Michigan, to a water purification plant. There, it is sand filtered using many more pumps to cycle it through the purification process. Pumps and more pumps, all of which are run on electricity. The electricity generated from a power house that burns fossil fuels. Plus, there is the required filtration process. In order for our water to be purified we had to construct purification buildings and produce the necessary chemicals used to ozonate, chlorinate, and fluorinate the water we drink. This is all made possible by some form of fossil fuels. Plus, I didn't even mention the infrastructure that was built and maintained, just to bring us the fresh water we drink. It doesn't take long and most people realize fossil fuels are an integral part of their everyday lives. At least at this point in time, most of the energy that sustains our population is supplied by fossil fuels. (Notes 91, page 191)

How is the inverted human pyramid holding up? In the industrialized nations it's holding its own. In many of the third world countries, the population size, and lack of food, is driving the pyramid to near collapse. In some cases, because of diseases, drought, wars, and political unrest, trophic levels have crashed. According to the United Nations, starvation and famine in parts of northeastern Africa have reached critical levels. (Besheer, 2017 April 24) Remember the words of Chief Seattle, "This we know…the Earth does not belong to us, we belong to the Earth… all things are connected…" (1854)

Can you believe it another song just came to mind! Hey man pass the word - be happy, find a copy of Pharrell Williams hit song, ♫ *Happy* ♫ (2014) and go play it. It's really "boogies!"

PYRAMID OF "E" HAS THE ANSWERS

THE FEEDING LEVELS ARE BASED ON "E" FLOW OR **CALORIES**

THE AMOUNT OF "E" NEEDED TO RAISE ONE GRAM OF WATER ONE DEGREE CELSIUS

AS TROPHIC LEVELS MOVE UP

1.) A LOSS OF AVAILABLE "E" OCCURS (2ND LAW)

RESULTS IN

2.) LESS AVAILABLE FOOD FOR ANIMALS

RESULTS IN A DECREASE IN POP.

3.) LESS FOOD MORE COMPETITION

THIS RESULTS IN LARGER INDIVIDUALS HAVING THE ADVANTAGE TO CATCHING THEIR PREY

NOTE: THERE ARE SMALLER POP.
OF LARGER ANIMALS FOR
THE SAME REASONS

LARGER INDIVIDUALS W/ SMALL
POP. AT THE TOP

"ENERGY DICTATES" W/ LARGER CRITTERS

DEER HAVE 4 LEGS?

— WHY EXPEND ENERGY ON HAVING
TOO MANY OFFSPRING:

* ONLY TO COMPETE FOR
  RESOURCES

* OR ONLY TO LOSE THEM
  TO STARVATION

IT'S FOR THE SAME REASONS MANY ANIMALS

* WALK IN THEIR STEPS → TO CONSERVE "E"
  IN THE SNOW OR MUD

* TAKE THE OPEN PATH TO THE WATER
  WHY CUT A NEW ONE? "E"

* HUMANS ARE NO DIFFERENT!

"THE PATH OF LEAST RESISTANCE"

HOW DO HUMANS/OUR POP. FIT INTO THE PYRAMID CONCEPT?

OVER POP.? vs. FOOD AVAILABILITY?

CREATING AN INVERTED PYRAMID

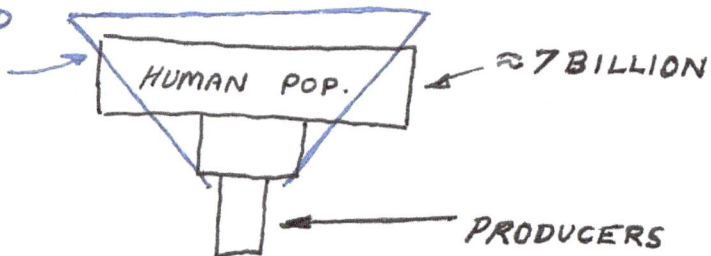

HUMAN POP. ≈ 7 BILLION

PRODUCERS

REMEMBER: THE PYRAMID'S SHAPE IS SET UP BECAUSE OF "E" FLOW IN THE ECOSYSTEMS

* W/ THE 1ST AND 2ND LAWS OF ENERGY

REMEMBER: WHAT THE HUMAN POPULATION IS DOING TO THE NATURAL PYRAMID CONCEPT

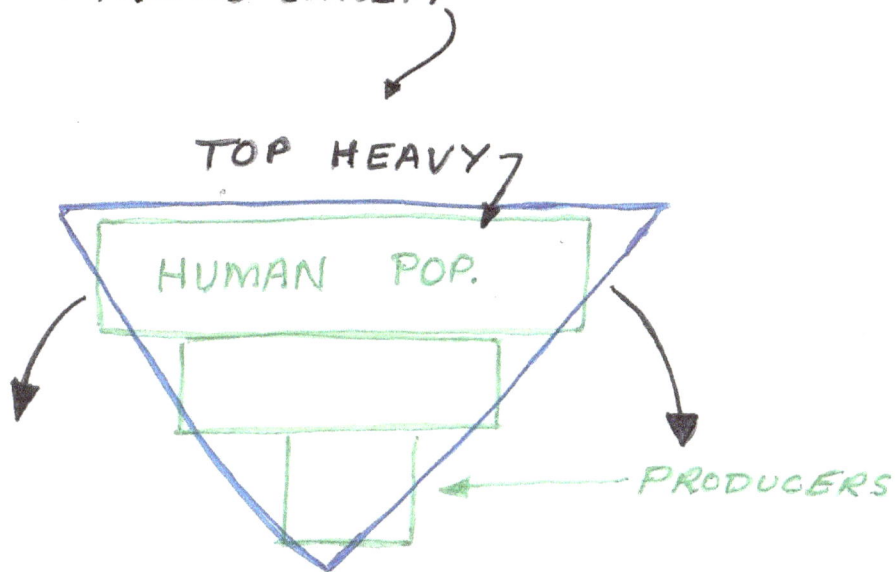

TOP HEAVY

HUMAN POP.

PRODUCERS

THE PYRAMID W/ HUMANS IN THE FOOD CHAIN

* IT'S BECOMING TOP HEAVY AND IS BEING INVERTED

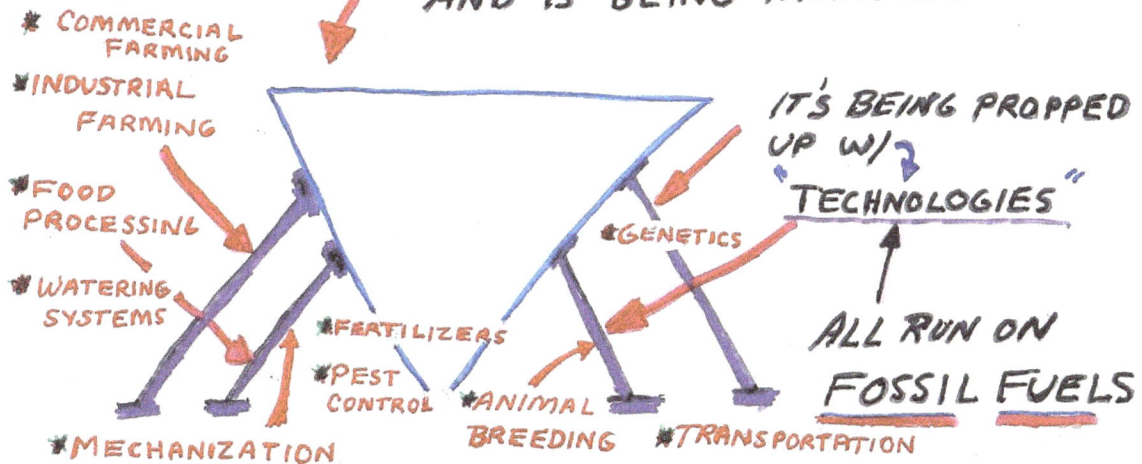

* COMMERCIAL FARMING

* INDUSTRIAL FARMING

* FOOD PROCESSING

* WATERING SYSTEMS

* FERTILIZERS

* PEST CONTROL

* MECHANIZATION

* ANIMAL BREEDING

* GENETICS

* TRANSPORTATION

IT'S BEING PROPPED UP W/ "TECHNOLOGIES"

ALL RUN ON FOSSIL FUELS

191

HUMAN IMPACTS ON THE NATURAL ORDER OF FOOD CHAINS AND FOOD WEBS

❋ IMPACTS CAN VARY DEPENDING ON THE

BIODIVERSITY OF A GIVEN AREA

LIFE VARIETY

→ THE NUMBER OF DIFFERENT SPECIES IN A GIVEN AREA

↑ BIODIVERSITY THE STRONGER THE ECOSYSTEM

BIODIVERSITY LEADS TO SOME ECOSYSTEMS DEPENDENT ON KEY SPECIES FOR THE SURVIVAL OF THE ECOSYSTEM

ALSO KNOWN AS

"KEYSTONE SPECIES"

Hi

OH NO

❋ LOSE THE KEYSTONE, LOSE THE WHOLE ECOSYSTEM

193

The human impacts on our natural world and its species have had, and continues to have, far reaching consequences on our biosphere. This is especially true when it comes to the world's population and its demand for clean air and water, food, energy, and other resources to sustain itself. In fact, this is an ecological topic that has been addressed in many research papers, articles, and books. I would like to remind you of the human impacts I addressed with each of my sections. I did this to reinforce the importance of knowing how our biosphere works naturally and how these fragile relationships are being unintentionally disrupted, causing many of the ecological problems we face today.

In this section I'd like to continue with that line of thought. Many food chains, given the biodiversity or varieties of life found in an area, will have food webs with multiple feeding options. (Notes 92, page 193) For example, many food webs have various food sources available if a particular predator, prey, or habitat is disrupted. However, in some cases it's just the opposite. There are some food chains and food webs that are very species specific, playing a key role in the survival of a whole ecosystem's structure. These are known as "Keystone Species." (Notes 93, page 195) The analogy I used with Keystone species was the bridge building techniques of the Romans. The Romans were very good at building aqueducts, bridges across waterways and/or canyons using the Keystone design. In fact, many of their engineering feats are still standing to this day. The Romans would begin by building scaffolding that arched across the area being bridged. (Notes 93, page 195) Next, stone cutters would lay stone against stone, working from both ends, fitting the stones tightly against one another until they met in the middle. At this point, a series of very specific, wedged-shaped stones were cut called "Keystones." The keystones were positioned to wedge the two sides of the bridge. Once these stones were installed, Roman engineers knew they could take down the scaffolding, because gravity pushed equally straight down. This force pushed the keystone and the surrounding stones, thus displacing the force in both directions of the bridge. From one stone to another, they would wedge each other to the very base on both sides of the bridge. This wedging process and displacing forces kept the bridge intact. (Brown, 2005) Consequently, if the Keystone is removed, the whole bridge would collapse.

KEYSTONE → BRIDGE BUILDING

ROCKS CUT AND
FIT TIGHTLY
WORKING FROM BOTH
SIDES

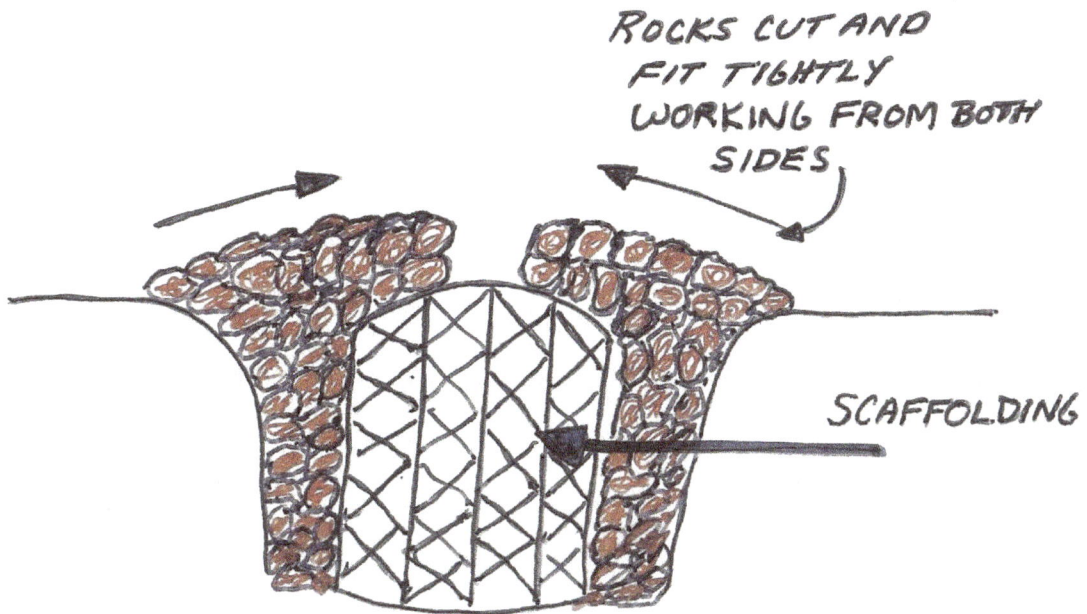

SCAFFOLDING

✻ THE WEDGING EFFECT IS DISPLACED TO OTHER ROCKS

"KEY STONE"
WEDGES ALL THE ROCKS

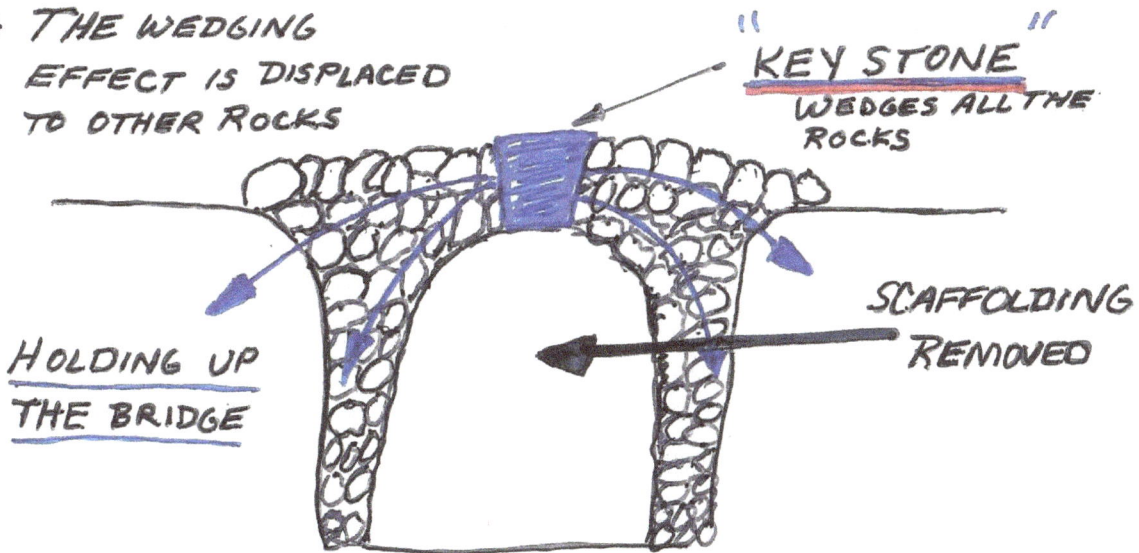

HOLDING UP THE BRIDGE

SCAFFOLDING REMOVED

✻ LOSE THE "KEYSTONE" AND THE BRIDGE COLLAPSES

Ex.) THE LOSS OF A TOP PREDATOR

"KEYSTONE" SPECIES → SEA OTTER VS. SEA URCHINS

W/O OTTERS, THE SEA URCHIN POPULATIONS INCREASED

↓

SEA URCHIN POPULATIONS DESTROY THE KELP FIELDS

↓

BREAKDOWN OF THE WHOLE ECOSYSTEM

Ex.) LOSS OF TOP PREDATOR → WOLVES IN YELLOWSTONE NATIONAL PARK

❋ HUNTED TO NEAR EXTINCTION
— TO PROTECT HUMANS AND LIVESTOCK

↓

ELK POPULATION DEVASTATED THE PLANT BIODIVERSITY — LOSS OF WOODY PLANTS/TREE SPECIES
— OTHER PLANT & ANIMAL COMMUNITIES

The same holds true in some ecosystems. If a Keystone species is lost, the whole ecosystem collapses. One of the well-known examples of this was the loss of the Sea Otter, which was a top predator. During the 1800's, the Sea Otter was hunted to near extinction. The Sea Otters fed on the Sea Urchins, keeping their populations in check. The Sea Urchin's meal of choice was the Kelp. Kelp is a large, leafy seaweed that grows like an underwater forest, supporting large varieties of species. Without the Sea Otters, the Sea Urchin populations grew out of control, which devastated the Kelp fields and brought down the whole ecosystem. (Arms, 2008) In other words, the Sea Otter was the keystone species. (Notes 94, page 196)

Another example of an ecosystem structure disruption was the loss of the wolf population in Yellowstone National Park. While it wasn't a loss of a keystone species, its impact on the surrounding ecosystem was detrimental to the biodiversity of the park (Notes 94, page 196 and Notes 95, page 198). Originally, the wolves were killed off because they were seen as a threat to human life and domesticated livestock. However, the wolf population played a significant role in keeping the elk herds in control because this was their most desired prey. Without the wolves, the elk population grew and their feeding ranges expanded. Elk were now free to graze on the open ranges consuming many of the woody trees that inadvertently affected many of the other plant and animal populations in the national park. Today, with the reintroduction of wolves into Yellowstone National Park, they have not only lowered the elk population but have also stopped the Elk's unchecked overgrazing of areas they normally wouldn't have moved into. (Stolzenburg, 2003 Fall) In turn, a vast variety of indigenous plant and animal species made a remarkable comeback strengthening the national parks natural biodiversity. Again, I would emphasize the importance of a top carnivore keeping many food webs in balance. (Notes 94, page 196 and Notes 95, page 198)

95

WITH THE REINTRODUCTION
OF WOLVES

↓

THE PATTERN WAS
REVERSED

ELK POPULATIONS
DECREASED

COYOTE POPULATIONS
DECREASED

WOODY TREE SPECIES
INCREASED

GRASS AND OTHER
HERBACEOUS PLANTZ
REBOUNDED

OTHER POPUATIONS OF
PRIMARY AND SECONDARY
CONSUMERS INCREASED

RESULTED IN

INCREASED
"BIODIVERSITY"

INCREASED THE HEALTH AND
SUSTAINABILITY OF THE
PARKS ECOSYSTEMS

Another predator-prey relationship disruption that has occurred began when a new, non-native, more aggressive organism is inadvertently introduced into an ecosystem. For example, an organism that is brought knowingly or unknowingly into an ecosystem that has no natural predator and is allowed to grow and reproduce unchecked. These organisms have become known as "Invasive Species." (Notes 96, page 202) Having no real predator, these organisms are allowed to flourish without regard to the ecosystem in which they were placed. As a result, these species outcompete the native species for resources causing the collapse of entire ecosystems. A case in point is the Zebra Mussel. (Notes 97, page 203) This is an invasive species that was inadvertently released from the bilge and ballast waters of ocean-going ships that came from the Caspian Sea region. They were originally released into the freshwater system of the Hudson River. From there, they have spread out making their way into the Great Lakes and beyond. The Zebra Mussel exemplifies an invasive that disrupts an ecosystem by stripping it of its resources. In fact, these guys are ferocious consumers. They filter feed at such a high rate that they outcompete the larger native species for much of the zooplankton and other microscopic organisms that are found in a freshwater ecosystem. In turn, this starves out the native populations of mussels, fish species, as well as other organisms that rely on these resources to survive.

Compounding the problem is most of the phytoplankton that provides oxygen to the freshwater ecosystems are also being consumed. This causes a depletion in the dissolved oxygen levels that sustain most of the aquatic species in a majority of fresh water ecosystems (Zebra Mussel Fact Sheet, n d).

Another significant impact humans are having on many ecosystems and their biodiversity is the cutting down of the rain forest. Here, whole ecosystems are being devastated because of population expansion and the need for more and more agricultural land. From South America to Indonesia, these forests are losing plant and animal species we don't even know about. Plus, the slash and burn practice of removing large sections of the world's rain forests are adding to the global greenhouse gas of carbon dioxide (CO_2). (Rawles, n.d.) As I mentioned earlier, the carbon cycle and fossil fuels act as a store of carbon, as do the Earth's forests of the earth, especially the rain forests. Instead of absorbing and storing carbon, the forests are now releasing it, adding to the global greenhouse effect, or global warming.

I would like you to think back to what Chief Seattle said, "All things are connected. . ." A great example of this concept is called "Biomagnification." It demonstrates how fragile these interconnections between the biotic and abiotic portion of our biosphere are that Chief Seattle spoke of. Biomagnification is the buildup or magnification of pollutants with each trophic level. (Note 98, page 204) (Campbell, 1996) In other words, harmful substances placed into our environment are retained in the tissues of living things. At each trophic level, these substances become more concentrated and toxic with each link in the food chain.

There are many documented cases of this process with long-lasting effects on ecosystems. One of the most well-known cases of Biomagnification was the use of DDT (Dichloro-Diphenyl-Trichloroethane), a pesticide now outlawed in most countries. DDT was an insecticide developed in the 1940s for military use to fight off insect-borne diseases. Because of its effectiveness on insect control, its uses expanded into farming practices and, institutional and home applications. With a wide range of use to control insects, it also worked its way into other animals in the food chain. What people didn't know was this effective insecticide was undermining many ecosystems and their predator-prey relationships. It was eventually discovered that DDT was very persistent in the environment, meaning it didn't break down. Researchers also found DDT to be soluble in lipids, or animal fats. It was in these fat tissues that DDT was being retained and built up over the course of an individual organism's life. (DDT, 2017 Aug.11) What started out at lower concentrations that didn't seem to have much impact on the lower order consumers, were now affecting individuals farther up the food chain. Remember, larger individuals need to eat more to meet their biological requirements. As individuals move up the food chain, they eat a bunch of little prey or one larger prey; in either case, they are consuming all of the DDT concentrations that the one or more individuals retained, or stored, in its entire life. When concentrations reach a critical level, it starts to adversely affect the health and reproductive systems of many top carnivores.

The effects of DDT were especially noticed in the Tertiary (3) and Quaternary (4) consumers. (Notes 81, page 172 and Notes 82, page 174) This was evident in the Bald Eagle population. (Notes 99, page 205) As a simple example how this process works, I would like you to think of an average fresh water lake or river. Now picture the abundance microscopic organisms, like Zooplankton and primary producers, growing and reproducing in that particular water system. I would also have you imagine DDT making its way, over time, into the lake from farm field runoff. Again, keeping this example in its simplest of terms, I would like you to think about all the microscopic organisms starting to absorb trace amounts of DDT over the course of their lives (Notes 98, page 204). Along comes a minnow that eats a lot of microscopic critters containing 1.0 parts per million (PPM) per individual organism. Over the course of the minnow's life, this one individual fish built up or retained a 100 PPM of DDT in its fat tissues. Then along comes a walleye, a much larger fish, and over the course of its life, eats 1000 minnows. The walleye has now picked up and retained 100,000 PPM of DDT in its lifespan. Next, a raptor such as an eagle comes down, catches and eats one walleye fish, and the eagle has now consumed 100,000 ppm DDT. (Notes 99, page 205) When the eagles would eat fish with high concentrations of DDT, it would either out right kill the adult eagle, or result in the loss of viable eggs to produce offspring. Researchers found that high concentrations of DDT interfered with the calcium production for the bird's eggshells. The eggs were weakened and were being crushed under the weight of mom or dad trying to incubate them. As a result, with few or no surviving eggs, the bald eagle population declined. It should be noted that once DDT was banned and with the help of recovery programs, the bald eagle has made a great comeback adding to the list of the many success stories of saving a once endangered species. (Campbell, 1996)

INVASIVE SPECIES

96

⌐→ IS AN ORGANISM THAT CAUSES
ECOLOGICAL OR ECONOMIC HARM
IN A NEW ENVIRONMENT WHERE
IT IS NOT NATIVE

<u>OR</u> ORGANISMS THAT ARE BROUGHT
INTO AN ENVIRONMENT CAUSING
ECOLOGICAL AND/OR ECONOMIC
HARM

* <u>NON-NATIVE</u>
- HAVE NO REAL PREDATOR
- CAN GROW AND REPRODUCE
  UNCHECKED
- OUTCOMPETES NATIVE
  SPECIES FOR RESOURCES

⬇

<u>RESULTS</u>
IN
⬇

THE DISRUPTION OF
THE ECOSYSTEM'S
PREDATOR-PREY
RELATIONSHIPS

"Look" **EXAMPLES OF INVASIVE SPECIES ARE NUMEROUS**   YIKES

I'LL USE THE <u>ZEBRA MUSSLE</u> AS AN EXAMPLE   Hi

<u>ZEBRA MUSSLE</u>
- ORIGINALLY FROM THE CASPIAN SEA REGION
- RELEASED INTO THE FRESH WATER SYSTEM OF THE HUDSON RIVER FROM THE BALLAST WATER OF A SHIP

Look

HAS NOW SPREAD INTO THE GREAT LAKES AND BEYOND

<u>WHAT IS IT?</u>
- BIVALVE THAT IS AN EXCEPTIONAL FILTER FEEDER

- BY FILTERING OUT MUCH OF MICROSCOPIC CRITTERS AS A FOOD SOURCE → OUTCOMPETES NATIVE SPICIES OF FOOD AND OTHER RESOURCES

BIOMAGNIFICATION

LIVING   INCREASE

- RETAINED SUBSTANCES
  BECOME MORE CONCENTRATED
  WITH EACH LINK IN A FOOD CHAIN

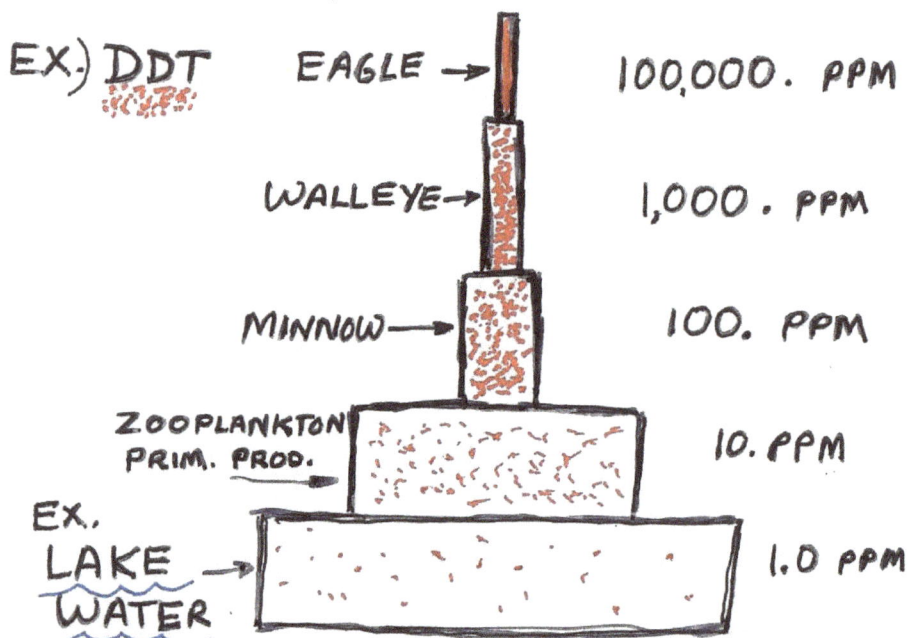

EX.) DDT

EAGLE →		100,000. PPM
WALLEYE →		1,000. PPM
MINNOW →		100. PPM
ZOOPLANKTON PRIM. PROD. →		10. PPM
EX. LAKE WATER →		1.0 PPM

DDT

- BANNED FROM USE IN MOST COUNTRIES
- WELL KNOWN PESTICIDE USED
  TO CONTROL INSECTS
- HIGH CONCENTRATIONS CAN
  KILL OTHER ANIMALS
✱ PERSISTS IN THE ENVIRONMENT
✱ SOLUBLE OR RETAINED IN LIPIDS
  ANIMAL FATS → CAN BUILD UP

Ex.)

BALD EAGLES AND DDT

EAGLES EAT FISH WITH
    HIGH CONCENTRATIONS OF
    DDT

↓

DDT INTERFERS WITH CALCIUM
PRODUCTION FOR EGG SHELLS

↓

EGGS BECOME
WEAK

↓

CRUSHED BY MOM OR DAD
    INCUBATING THE EGGS

↓

DECLINE
OF EAGLE POPULATIONS

"COOL"   "LOOK"

WITH THE BAN ON DDT AND
RECOVERY PROGRAMS, THE
EAGLE POPULATIONS HAVE
REBOUNDED

99

I must remind you these were just a few of the many examples of the impacts man has had on our ecosystems and predator-prey relationships. Also, by knowing how our environment works, we can make great strides in correcting past practices and plan a future that works closer with nature and not against it. Remember, "...We didn't weave the web of life, we were merely strands in it...Whatever we do to the web, we do to ourselves..." (Chief Seattle, 1854)

To end this ecological writing may I suggest another song, - I do, I do, I must, I must, Danka Schoen by Wayne Newton (1963) - it's so choice! Thank you very much!

**Now buzz on over
to Section Five!**

SECTION

FIVE

Author's Notes

I wrote this book to give an average person a brief but important look into how our Ecology works. It's by no means all inclusive. It's another way to put forth one's ideas and/or concerns for others to ponder and to gain a general understanding of our home, Earth, and its importance. In fact, the study of our home is a matter of our survival. We are no different than any other critter that inhabits this planet. The more we know, the better our chances. If man's ability to learn and reason sets us apart from other species, then survival of the fittest will dictate whether or not we, as a species, will survive. What we have come to know cannot be ignored. We need to protect ourselves from our capacity for ignorance. What's more important for our survival than clean air and water, wild lands and fertile soils? What has greater long-term value than the understanding and respect for the natural mechanisms that has allowed life to exist long before man's arrival?

If progress means profit (short term value), what will be the long-term costs for future generations or all of humanity itself? How much is enough? We need to take note and save ourselves from moving backwards and, instead, move forward from what we've learned from past events. What would it be like if there were no more wild places? Or, where wildlife is relatively free of man's impacts? Wallace Stegner once said, "We need wild country... even if we never do more than drive to its edge and look in. For it be... a part of the geography of hope." (Stegner, 1971)

Remember, "We are all connected... all things are connected..." as profoundly stated by Chief Seattle, 1854.

In honor and remembrance of all of life's interconnections, I made my own personalized symbol that represents our Ecology. I paced it at the beginning of this section. I love every personalized thought in its making!

Here is the meaning of my symbol

THE TRIANGLE

All things are connected
All things come in threes
The Mobius Strip for recycling
Three dimensions
Three ingredients for fire - O_2, Activation E, fuel to burn
Three points of personal wisdom: lawful, moral, ethical

THE SUN

Round shape – the circle of life
Source of all E
Allows life to exist
Smiley Face - To look for humor in life
Six Rays – The six most important elements
Six divided by two equals three
O, H, C, N, Si, Fe
Personal warmth

THE SUNFLOWER

Represents the Land – One of the three parts of the biosphere
3 Roots and 3 Leaves
Four petals – for the diversity of life
Smiley Face - Represents the beauty of nature
From the soil life abounds, only to return
The Flow of E to other life forms

THE THREE BIRDS

Represents the Atmosphere – One of the three parts of the biosphere
Connects the land and water ecosystems
Regulates the Earth's temperature
Protects the Earth from harmful solar rays – O_3
Life takes flight
Carries the gases of nutrients, E, and life

THE WAVES OF WATER

Represents the Water – One of the three parts of the biosphere
Water is the essence of life
Water has unique chemical properties
Water, especially our oceans, is the key to life on our planet
Regulates our planet's temperature
Takes in and releases important gases – $O_2 N_2$, CO_2

"…and that's what it's all about Charlie Brown"

BIBLIOGRAPHY

11 Countries Leading the Charge on Renewable Energy. (2019, Jan. 13). Climate Council. Retrieved from
    https://www.climatecouncil.org.au/11-countries-leading-the-charge-on-renewable-energy/

Andrews, W.A. (1972). Environmental Pollution: A Guide to the Study of Environmental Pollution.
    Englewood Cliffs, NJ: Prentice-Hall Inc.

Arms, K. (2008). Environmental Science. Austin TX: Holt, Rinehart and Winston.

Barris, C. (Producer and Director). (1976). Gong Show. Retrieved from   https://www.imdb.com/title/tt0133303/

Bellis, M. (2018, March 19). How Tidal Power Works: There are Three Ways We Can Harness
    Tidal Power Thoughtco. Retrieved from https://www.thoughtco.com/how-tidal-power-plants-
    work-1992544

Benarde, M.A. (1973). Air Pollution. Our Precarious Habitat: An Integrated Approach to Understanding
    Man's Effect on His Environment. New York, NY: W.W. Norton and Company.

Berg, L. (2017, April 24). The Average Height of Redwood Trees. Science. Retrieved from
    https://sciencing.com/average-height-redwood-trees-6086324.html

Besheer, M. (2017, March 10). UN Aid Chief: 20 Million People in 4 Countries Face Starvation, Famine.
    Retrieved from https://www.voanews.com/a/twenty-million-people-four-countries-face-starvation-fam
    ine-un-aid-chief-says/3760816.html

Brown, D.J. (2005). Bridges: A Thousand Years of Defying Nature. Buffalo, NY: Firefly Books.

Bunker, A. (2018, Nov. 18). Peromyscus maniculatus: Deer Mouse. Animal Diversity Web. Retrieved from
    https://animaldiversity.org/accounts/Peromyscus_maniculatus/

Campbell, N.A. (1996). Biology (4th ed.). Riverside, CA: The Benjamin/Cummings Publishing Company, INC.

Chief Seattle. (1854). Chief Seattle's letter to all.
    Retrieved from http:www.csun.edu/ ⊠ vcpsy00/seattle.htm.

Davenport, C. (2018, Oct. 7). Major Climate Report, Describes a Strong Risk of Crisis as Early as 2040.
    Retrieved from https://www.nytimes.com/2018/10/07/climate/ipcc-climate-report-2040.html

DDT – A Brief History and Status. (2017, Aug. 11). United States Environmental Protection Agency.
    Retrieved from https://www.epa.gov/ingredients-used-pesticide-products/ddt-brief-history-and-status

Diamond, J. (1997). Guns, Germs, and Steel. New York NY: W.W. Norton and Company.

Ecological Pyramids. (n.d.). [PDF Document]. Retrieved from
    https://biology.tutorvista.com/ecology/ecological-pyramids.html

E.P.A. (2016). International Actions-The Montreal Protocol on Substances that Deplete the Ozone Layer.
    Retrieved from https://www.epa.gov/ozone-layer-protection/international-actions-montreal-protocol-
    substances-deplete-ozone-layer

Eunson, S.J. [PBS]. (1990, Jan. 1). PBS: Race to Save the Planet Series. Part 2: Only One Atmosphere.
    Retrieved from https://www.youtube.com/results?search_query=PBS+race+to+save+the+planet+series%
    2Conly+one+atmosphere

Faith. (2017). The Bread Basket of the World. On world Atlas. Com. Retrieved from
    https://www.worldatlas.com/articles/thebreadbaskets-of-the-world.html

Fox, D.L. (2007). Vulpes vulpes: Red Fox. Animal Diversity Web. Retrieved from
https://animaldiversity.org/accounts/Vulpes_vulpes/

Henn, C. (2016, Sept. 12). The growing Problem of Toxic Blue Green Algae Blooms (Cyanobacteria).
Surfrider Foundation. Retrieved from https://www.surfrider.org/coastal-blog/entry/the-growing-prob
lem-of-toxic-blue-green-algae-blooms-cyanobacteria

Hindenburg. (2018). History. Com Editors. Retrieved from https://www.history.com/topic/greatdepression/Hin
denburg

Hydro Plants. (2018). Wisconsin Valley Improvement Company. Retrieved from
http://www.wvic.com/Content/Hydroplants.cfm

Lallanilla, M. (2013, Sept. 25). Chernobyl: Facts about the nuclear Disaster. Live Science. Retrieved from
https://www.livescience.com/39961-chernobyl.html

Leopold, A. (1949). A Sand County Almanac and Sketches Here and There. New York, NY: Oxford
University Press.

Meyer, R. (2018, Apr. 18). Since 2016, Half of all Coral in the Great Barrier Reef Has Died. Science.
Retrieved from https://www.theatlantic.com/science/archive/2018/04/since-2016-half-the-coral-in-the-
great-barrier-reef-has-perished/558302/

Mobius. (n.d.). Bookrages.com. Retrieved from http://www.bookrages.com/biography/august-augustus-
ferdinand-mobius-wom/

Oram, P. G. (n.d.). Dissolved Oxygen in Water. Water Research Center. Retrieved from
https://www.water-research.net/index.php/dissovled-oxygen-in-water

Parker, L. (2018, June). Plastic. National Geographic, 40 – 91.

Petrucci, R.H. (1989). General Chemistry, Principles and Modern Applications (5th ed.). New York, NY:
Macmillan Publishing Company.

Piraro, D. Bizarro.com. Distributed by King Features.

Rawles, S. (n.d). Forest Habitat. Wildlife found. Retrieved from
https://www.worldwildlife.org/habitats/forest-habitat

Rifkin, J. (2002). The Hydrogen Economy. New York NY: Jeremy P. Tarcher/Putnam.

Sagan, C. (1994). The Pale Blue Dote. New York, NY: The Random House Publishing Group.

Samels, M. (Producer and Director). (2006). Hoover Dam. [DVD]. United States: WGBH Educational
Foundation.

Seattle, C., Jeffers, S. (1991). Brother Eagle Sister Sky. New York, NY: Puffin Books.

Stegner, W. (1971). Wallace Stegner Quotes (Author of Angle of Repose). Good Reads quotes. Retrieved
from https://www.goodreads.com/author/quotes/157779.Wallace_Stegner?page=6

Stolzenburg, W. (2003, Fall). The Long Rangers. Nature Conservancy, 35 – 43.

Three Mile Island Accident. (2012, Jan.). Nuclear Energy Institute. Retrieved from
http://www.world-nuclear.org/information-library/safety-and-security/safety-of-plants/three-mile-
island-accident.aspx

Velikovsky, I. (1950). Worlds in Collision. Garden City, NY: Double Day and Company, Inc.

Vitale, A. (n.d.). African Bush Elephant. The Nature Conservancy. Retrieved from https://www.nature.org/en-us/explore/animals-we-protect/african-bush-elephant/

Westerling, A.L., Hidalgo, H.G., Cayan, D.R. Swetnam, T.W. (2006, Aug. 18). Warming and Early Spring Increase Western U.S. Forest Fire Activity. Science, 313. Retrieved from http://science.sciencemag.org/content/313/5789/940

Why Aluminum Cans? (n.d.). Canned Water 4 Kids. Retrieved from http://www.cannedwater4kids.com/index.php?/why-

World Population: July 11,2018. (2018, July, 11). U.S. Census Bureau. Retrieved from https://www.census.gov/newsroom/stories/2018/world-population.html

Zebra Mussel Fact Sheet. (n.d.). Cary Institute of Ecological Studies. Retrieved from https://www.caryinstitute.org/sites/default/files/public/downloads/curriculum-project/zebra_mussel_fact_sheet.pdf

Songs: Listed generally in the Order of their Appearance

Pauling, L., Bass, R. (1967). Dedicated to the One I Love. [The Mamas and The Papas].

John, E., Rice, T. (1994). Circle of Life. [John, Elton].

Lyle, G.H., Britten, T. (1984). What's Love Got to Do with It. [Turner, Tina].

Gaye, G., Gaye, G.P. (1971). Mercy, Mercy Me, The Ecology. [Gaye, Marvin].

Livgren, K. (1977). Dust in The Wind. [Kansas].

Morricone, E. (1967). The Good, The Bad, and The Ugly. [Nicolai Orchestra].

Scott, A., Griffin, T. (1978). Love is Like Oxygen. [Sweet].

Lennon, J., McCartney, P. (1968). Helter Skelter. [Beatles].

Wainwright, L. III (1972). Dead Skunk in the Middle of the Road. [Wainright, Loundon III]

Dean, J., Acuff, R. (1961). Big John. [Dean, Jimmy].

Mitchell, J. (1967) Both Sides Now. [Collins, Judy].

Clapton, E. (1974). Let It Grow. [Clapton, Eric].

Ramsey, W.A. (1976). Muskrat Love. [Captain and Tennille].

Larkin, J., Nunzio-Catania, A. (1995). Scatman (Ski-ba-bop-ba-dop-bop). [Scatman John]

Dylan, R. (1971). If Not for You. [Newton-John, Olivia].

Mitchell, J. (1970). Big Yellow Taxi. [Mitchell, Joni].

Mitchell, J. (2002). Big Yellow Taxi. [Counting Cows and Vanessa Carlton].

Bono, S. (1967). The Beat Goes On. [Sonny and Cher].

Jolicoeur, D., Mason, V., Mercer, K., Huston, P. (1961). A Little Bit of Soap. [The Jarmels].

Nielsen, R. (1979). Dream Police. [Cheap Trick].

Kemp, W. (1976). One Piece at a Time. [Cash, Johnny].

Fischoff, G., Powers, T. (1967). 98.6. [Keith].

Dylan, R., Harrison, G., Lynne, J., Orbison, R.K., Petty, T.E. (1988). End of the Line. [Traveling Willburys].

Willis, A., Semibello, D. (1983). Neutron Dance. [Pointer Sisters].

Grant, E. (1982). Electric Avenue. [Grant, Eddie].

Williams, P. (2014). Happy. [Williams, Pharrell].

Kaempfert, B., Schabach, K., Gabler, M. (1963). Danka Schoen. [Newton, Wayne].

Movies: Listed Generally, in the Order of Their Appearance

Leyroy, M. (Producer), Fleming, V. (Director). (1939). Wizard of Oz. [DVD]. United States: Metro-Goldwyn-Mayer.

Warner, A., Williams, J.H., Katzenberg, J. (Producer), Jenson, V. (Director). (2001). Shrek. [DVD]. United States: Dream works Animation.

Bergman, A. (Producer), Brooks, M. (Director). (1974). Blazing Saddles. [DVD]. United States: Warner Brothers Pictures.

Melledandri, C., Healy, J., Cohen, J. (Producer), Renaud, C., Coffin, P. (Director). (2009). Despicable Me. [DVD]. United States: Universal Pictures.

The Actors Home Episode. (1953). Abbot and Costello, Who's on First. Retrieved from https://www.youtube.com/watch?v=kTcRRaXV-fg

Godon, L., Silver, J., Davis, J. (Producer), McTiernan, T. (Director). (1987). Predator. [DVD]. United States: Twentieth Century Fox.

Lazar, A., Roven, C., Gartner, A., Ewig, M. (Producer), (Director). (2008). Get Smart. [DVD]. United States: Warner Brothers Pictures.

Grazer, B. (Producer), Howard, R. (Director). (2005). Apollo 13. [DVD]. United States: Universal Pictures.

Hughes, J., Jacobson, T. (Producer), Hughes, J. (Director). (1986). Ferris Bueller's Day Off. [DVD]. United States: Paramount Pictures.

Mendelson, L., Melendez, B. (Producer), Melendez, B. (Director). (1965). A Charlie Brown Christmas. [DVD]. United States: Warner Brothers Pictures.

Quick Index to notes, topics, and their pages